# Pragmatic Justifications for the Sustainable City

What can justice and sustainability mean, pragmatically speaking, in today's cities? Can justice be the basis on which the practices of city building rely? Can this recognition constitute sustainability in city building, from a pragmatic perspective? Today, we are faced with a mountain of reasons to lose hope in any prospect of moving closer to justice and sustainability from our present position in civilization.

*Pragmatic Justifications for the Sustainable City: Acting in the Common Place* offers a critical and philosophical approach to revaluating the way in which we think and talk about the "sustainable city" to ensure that we neither lose the thread of our urban history, nor the means to live well amidst diversity of all kinds. By building and rebuilding better habits of urban thinking, this book promotes the reconstruction of moral thinking, paving the way for a new urban sustainability model of justice.

Utilizing multidisciplinary case studies and building upon anti-foundationalist principles, this book offers a pragmatic interpretation of sustainable development concepts within our emerging global urban context and will be a valuable resource for both undergraduate and postgraduate students, as well as academics and professionals in the areas of urban and planning policy, sociology, and urban and environmental geography.

**Meg Holden** is Associate Professor in the Urban Studies Program and Department of Geography, Simon Fraser University, Vancouver, Canada.

# Routledge Equity, Justice and the Sustainable City series

Series editors: Julian Agyeman, Zarina Patel, AbdouMaliq Simone and Stephen Zavestoski

This series positions equity and justice as central elements of the transition toward sustainable cities. The series introduces critical perspectives and new approaches to the practice and theory of urban planning and policy that ask how the world's cities can become 'greener' while becoming more fair, equitable and just.

Routledge Equity, Justice and the Sustainable City series addresses sustainable city trends in the global North and South and investigates them for their potential to ensure a transition to urban sustainability that is equitable and just for all. These trends include municipal climate action plans; resource scarcity as tipping points into a vortex of urban dysfunction; inclusive urbanization; "complete streets" as a tool for realizing more "livable cities"; the use of information and analytics toward the creation of "smart cities".

The series welcomes submissions for high-level cutting edge research books that push thinking about sustainability, cities, justice and equity in new directions by challenging current conceptualizations and developing new ones. The series offers theoretical, methodological, and empirical advances that can be used by professionals and as supplementary reading in courses in urban geography, urban sociology, urban policy, environment and sustainability, development studies, planning, and a wide range of academic disciplines.

**Incomplete Streets**
Processes, practices and possibilities
*Edited by Stephen Zavestoski and Julian Agyeman*

**Planning Sustainable Cities and Regions**
Towards more equitable development
*Karen Chapple*

**The Urban Struggle for Economic, Environmental and Social Justice**
Deepening their roots
*Malo Hutson*

**Bicycle Justice and Urban Transformation**
Biking for all?
*Edited by Aaron Golub, Melody L. Hoffmann, Adonia E. Lugo and Gerardo Sandoval*

**Green Gentrification**
Urban sustainability and the struggle for environmental justice
*Kenneth Gould and Tammy Lewis*

**Pragmatic Justifications for the Sustainable City**
Acting in the Common Place
*Meg Holden*

# Pragmatic Justifications for the Sustainable City

Acting in the Common Place

**Meg Holden**

First published 2017 by Routledge

2 Park Square, Milton Park, Abingdon, Oxfordshire OX14 4RN

52 Vanderbilt Avenue, New York, NY 10017

*Routledge is an imprint of the Taylor & Francis Group, an informa business*

First issued in paperback 2018

*British Library Cataloguing in Publication Data*
A catalogue record for this book is available from the British Library

*Library of Congress Cataloging in Publication Data*
Names: Holden, Meg, author.
Title: Pragmatic justifications for the sustainable city : action in the common place / Meg Holden.
Description: Abingdon, Oxon ; New York, NY : Routledge, [2017] | Series: Routledge equity, justice and the sustainable city | Includes bibliographical references
Identifiers: LCCN 2016042581| ISBN 9781138121102 (hbk) | ISBN 9781315651255 (ebk)
Subjects: LCSH: Sustainable urban development. | City planning–Environmental aspects. | Urban ecology (Sociology)
Classification: LCC HT241 .H65 2017 | DDC 307.1/16–dc23
LC record available at https://lccn.loc.gov/2016042581

ISBN: 978-1-138-12110-2 (hbk)
ISBN: 978-0-367-15222-2 (pbk)

Typeset in Bembo
by Taylor & Francis Books

# Contents

"Drawing on the American pragmatic tradition and the recent pragmatic French sociological theory, Meg Holden develops a fresh and illuminating approach to issues of urban sustainability and justice. She perceptively discusses recent debates and persuasively shows how a pragmatic orientation provides a more flexible and realistic way of moving forward with urban planning. Throughout she shows a subtle way of integrating theory and practice."

*Richard J. Bernstein, New School for Social Research, USA*

# Figures

# Preface

In April 1991, I filed in line to board a yellow school bus to a rival high school's gymnasium for a county-wide Earth Day event. The event had been planned to engage me and my fellow teenagers on the 20-year anniversary of Earth Day and included lots of milling about, awkward giggling and oogling, information booths and stickers, some political speeches (Ontario premier-to-be Bob Rae was there), and Dr. David Suzuki. I remember that attendance was optional, and at that time in my so-far suburban life playing hookey with friends was a much more popular thing to do than attending planned rallies. To be truthful, doing that which my friends were doing was the default decision. This time, though, I got on the bus although I don't recall any of my friends opting to go. Tony was there, my school's star basketball player and my first crush. He was unique in my mind not for his fame but because he told me in one of our rare conversations that his favourite subject was geography. Conditioned by peer pressure, I didn't understand at the time how the subject of geography could in any way achieve such status in the mind of a status-concerned kid. Anyway, we may have said hi, but we didn't talk at the Earth Day rally.

The event was packed with students who I guess had also overcome their self-consciousness about feeling uncool for caring about Earth Day when they could be enjoying time away from school, or maybe they were more socially advanced than I was. (I was 25 before I figured out that the coolness rules on which I had based my understanding of my own self-worth were complete bullshit. Coincidentally, I was 25 when I moved to New York City.) I don't remember talking to any of my fellow students. It's possible I simply couldn't, because I do remember being struck speechless by the crisis, the challenge, and the marching orders handed down to me that day, particularly by Suzuki. He unceremoniously and straightforwardly pulled the cord on the blinders in front of my blue eyes about the condition of the world, the position of my country in it, the work that lay before the willing, and the implications of not getting personally involved, deeply, utterly, and with gusto.

Up until this time, I didn't think anything could be more moving than a Pink Floyd song. I was already a self-proclaimed environmentalist, contributing money earned from babysitting to both Greenpeace and the World Wildlife

Fund, but all things considered, this was mostly about the bumper stickers these organizations sent me back, which I used to decorate my school binders. Hearing Suzuki's presentation on the perils of the planet sent me over the deep end. Whales and baby seals were being killed, rainforest destroyed, toxic chemicals released from my paper and most of the products I used, and the atmosphere was being seriously messed with; all this I already knew. But he convinced me that I was implicated, my family was implicated, and in my ignorance of this I had already been shirking my responsibility while the dire situation deteriorated.

I went home and did what Suzuki said to do: I wrote signs to my parents and posted them on my bedroom door, on the refrigerator, and even on their bathroom mirror. Following his careful instructions, the signs said: EVERY 6 SECONDS, 133,000 SPECIES ARE GOING EXTINCT, and WE HAVE 10 YEARS TO CHANGE THE WORLD and WOULD YOU IGNORE YOUR DAUGHTER IF SHE TOLD YOU SHE HAD CANCER?

For the record, this initiation to home-based direct action on my part had exactly zero impact. Not one conversation, or changed idea, or diverted or reallocated shopping dollar, or altered habit resulted in my household, present writer excluded. David Suzuki would not be proud of me, I thought, and decided not to take activist ideas any further. I obviously wasn't very good at it and if Suzuki ever found out, I would be humiliated. I was pretty shy, besides. I did have a knack for formal learning, however, and so this was an obvious place to invest my energy for what became, the following year at Rio, the quest for sustainable development. Equipped with the confidence of 20 more years of thinking about sustainable development, the credibility of a PhD, and experience as an urban researcher engaged with policy, activism, and organizations in a host of cities, this book demonstrates how I am coming to terms with the sorry results of my early attempts at activism and making some ginger steps back out "to the street" with my ideas. My mother's home, of course, remains the final frontier.

This is a book about hope for and in the city to become a place for what I don't believe we have ever before seen on Earth: sustainability. The approach taken is philosophical investigation of how a pragmatic interpretation of sustainable development concepts, plans and policies articulates a path toward sustainable cities. Amidst the intellectual and policy furore over defining "the sustainable city," there may be infinite variations on the theme of constructing and maintaining the urban realm in a manner fit to promote better human development and livelihood opportunities, long into the future, without depleting their natural resource base or decreasing life opportunities for those elsewhere around the world. Building upon pragmatic, anti-foundationalist principles, we will attempt here to build a definition from the ground up. We have over 200 years of history in democratic society from which, if we can face the challenge of dissecting the hope and promise from the hardship and failure, we can begin to build a sustainable city that is within the realm of possibility. At any rate, from a pragmatic perspective, this history, along with the

Financial support came from the Canadian Social Sciences and Humanities Research Council Insight grant 435-2014-0465 as well as from a National Research Foundation of Korea Grant funded by the Korean Government (NRF-2013S1A3A2054622).

Thank you for reading.

# Part I

# Our starting point for urban sustainability and justice

# 1 Our starting point

## Sustainability and justice made urban

What can justice and sustainability mean, pragmatically speaking, in today's cities? Can justice be the basis on which the practices of city building rely? Can this recognition constitute sustainability in city building, from a pragmatic perspective? Today, we are faced with a mountain of reasons to lose hope in any prospect of moving closer to justice and sustainability from our present position in civilization. But from the philosophically pragmatic approach that we adopt in this book, despair is not an option. Instead, what we attempt here is to take stock of the demand for sustainability and justice within a growing awareness of the globally dominant trends of urbanization. We will consider the critical urban scholarship that warns us about the ways in which finding hope in contemporary urban trends is dangerous. Specifically, critical urban scholarship warns us about three traps in the contemporary celebration of prospects for justice and sustainability in cities: (1) the local trap; (2) the empowerment trap; and (3) the community trap.

The warnings are well advised, but we will consider how bringing the resources of pragmatic thinking to bear on the trappings of urbanism today can change the equation. With a mix of reasoning from philosophical, sociological, and urban studies bases, cases and anecdotes, we will propose that some key reasons for hope sit with urbanism today. We build a case, in short, that some uniquely urban values can take root today, offering promise to move us toward sustainability justice.

First, urbanites today give value to an individualist authenticity that, for all its vanity and obsessive qualities, also opens up a willingness to engage with diverse others. Because contemporary urbanity values an individualist sense of self-determination and life planning, radically uprooted from any sense of given foundations, urbanites also have the potential to develop better habits of tolerance of diverse lifestyles and perspectives.

Second, today's successful cities may mock the prospect of any quintessential utopia devised from above, in advance, but urbanites have not given up on crafting their own versions of utopia. In generating a willingness among urbanites to engage with one another in piecemeal ways to create partial, experimental, fleeting utopian projects and alternatives, the city includes a pragmatic utopian vision that also serves to empower those who engage in crafting and carrying out these experiments.

Third, and finally, the appeal of the city today raises the spectre of risk as a value, that offers rewards to those willing to engage. Demonstrating resilience in the face of risk, rather than simply seeking security, suggests an opening-up of options to consider new possibilities, new prospects, new arrangements to favour community worth and wellbeing beyond market determinations.

There is no question that the celebration of the city for its creativity, its flexibility, its potential to mix people together and save resources and energy as never before, far out-reaches what cities have actually delivered in these respects. At the same time, in this book, we will argue from a pragmatic perspective that these failed expectations need not be the end of the story. Together, the move toward a uniquely urban authenticity, the move toward a level of engagement and local experimentation in common and in public space, and the move toward an embrace of resilience that rewards the risk of rearranging our positions and perspectives, constitute contemporary urban values that may yet be put to the service of justice and sustainability.

The new notion of urban justice that I craft in this book draws from the philosophy of American pragmatism, principally that of John Dewey, William James, Charles S. Peirce, and Jane Addams, as well as the French school of pragmatism of engagement based at l'Ecole des Hautes Etudes en Science Sociales, with Laurent Thévenot and Luc Boltanski. In very brief outline, the pragmatic construct of justice that I will offer is based upon a morality of engagement with the city, and drawing out a series of different modes of engagement with the city in public as well as quasi-public, familiar spaces. Thévenot considers it a mistake to distinguish the meaning of the French "justice" (justice) and "justesse" (aptness, appropriateness) – following the same line of thinking, we will present an understanding of how "justification" can be approached as another way of thinking of the action that is demanded of us if we are to move in the direction of "justice."

In contemporary cities, which by their very nature are open to a diversity of perspectives, lifestyles, and choices, we can start by considering an ordinary sense of justice. We can begin with a basic sense of urban democracy, of an ethic of civic and moral association in social spaces. The justice and sustainability that we are bound to try and generate today are plural and demand an articulation in community, in the public sphere, via action in common that depends crucially on context but also crucially upon habit. This understanding of justice as engagement and justification within a distinctly urban context of diversity and democratic as well as sustainability aspirations, can be argued from a pragmatic basis.

To begin our search for the city that is acting in the direction of sustainability and justice, we start with the ordinary sense of justice and the populist sense of sustainability. We start with the exercise of a "mental fly-over" or an imaginary omniscient gaze upon a city in a state of exception. We pretend for a moment to have a fulsome view of the diversity of perspectives on justice and sustainability in a city in which governmental and behavioural norms have been called off, for a brief time. Our starting point, then, is Quebec City, a fortified city, once built to ensure security for those within from threats that came from beyond. Entry to the old city of Quebec is limited to a winding, reinforced road up the steep hillside, which is also icy for the long winter, which lasts from November to April, sometimes longer. We will imagine the streets of this old city filled with seekers of justice and sustainability, because the city is hosting a meeting of world leaders to make joint progress toward these goals. We will not concern ourselves with the formal political endeavours in which the leaders themselves are engaging. We put this work to one side, and instead look to the streets, mobilized because of the impetus of this meeting toward an ordinary sense of justice and populist sense of sustainability. Just what is it that this community can be considered to be working on, and how?

## The city in a state of exception

On the day of our observation of the social life and justifications operating in this city, the city is in a state of exception. The city is host, we will imagine, to heads of state, leaders of international organizations, and civil society organizations, and the unaffiliated justice-seekers and sustainability-seekers of the world for an unprecedented meeting of the Forum for Progress and World Future. Quebec City has been transformed from its picturesque, simulacrum of a fortified town of older, forgotten times and stands up now to its old functional fortified structure, with controlled and narrow entry and exit (some would say that it was chosen as host city for this very reason). On this first day of the Forum, what are intended to be three days of intensive multilateral meetings among an unprecedented, flat network-of-networks of heads of state, heads of international economic and nongovernmental organizations, heads of city and regional governments, and a parallel meeting of civil society organizations, the justice-seekers of the world − those with the means and the ability to travel − have converged on the narrow, cobbled streets of Old Quebec to express the need for change, as they see it, and as they see fit to advance it.

The day begins with a large public march up the Rue de la Couronne to Old Quebec. The delegates continue from here into the official meeting halls to begin formal negotiations. There are seven meeting halls operating on the first day, with one allocated for each continent (subsequent days will organize the different meeting halls thematically − the second day, and generationally − the third and final day). We, however, will stay in the street, where many thousands of people hungry for progress and a view to world future remain, representing a wide range of agendas from the personal to the global, and advancing them via every route imaginable, from offering services, to selling products, to performing, to engaging in dialogue and other information-sharing activities.

Beginning near the base of the hill climb, early in the day, there is a tent. It is full of people, of many ages, many appearances, and many languages. Tarps, sleeping bags and reusable water containers are piled up in one corner. In another sits a first aid station bearing a red cross, a red crescent, and a red Om. First aid attendants are offering to add vinegar to anyone's water bottle to counteract the tear gas, for later on. A bank of interpreters is seated along the back of the tent, whispering into their headphones. In addition to headphones for the speakers on stage, there are secure mobile communication devices to loan, quantum encrypted, with a piece of collateral taken as deposit. A woman with glossy dark hair and a black leather jacket is the master of ceremonies. She is flanked by a French Canadian comedian wearing a capitalist clown outfit. They are talking about the merits of forming affinity groups at this early stage in the protest event.

A murmur of disgust can be heard from a young person, in a black hood: "We are not acting in partnership! We are militants, in our streets that belong to us!"

Some of the crowd is listening to what is happening on stage, some are pamphleting, some caucusing in small groups, helping one another with masks and bandanas, some are preparing signs and banners, some are bursting into song –

L'eau est la vie et la vie n'a pas de prix! [Water is life and life has no price!]

Perhaps connected to this song, perhaps from a different starting point altogether, a long silk river dragon puppet winds its way through the tent, a centipede of feet underneath supporting it, making sounds like running water and continuing its journey outside.

We slip out of the tent too, and begin our climb. Stations and groupings of people line the way, some seated cross-legged, some standing, some snaking their way up the hill along with us. The crowd is gathering, the sense of energy coalescing is palpable. Bicycle helmets and thermoses and all manner of other objects are serving as drums. A beautiful woman with very long hair and feathers drums slowly and sings as if she might never have to stop – "If my body swallowed my tears; if my body swallowed my blood." *Ad infinitum.*

Some boutiques, shops, and cafés remain open, their curious, enthusiastic, or concerned shop keepers in the doorway, observing the scene as it unfolds. Some are taking deliveries of boards to put up over the shop windows. Young families, small groups of hapless tourists, and a few Quebecois in fashionable clothes, out looking for cocktails or cigarettes, are slipping in and out of the shops, still. A young woman in black walks by them, carrying a hand-made sign that says: "Citoyen ou esclave? Par où la démocratie?" [Citizen or slave? Which way is democracy?]

One hungry local gets turned away from his favourite café, in the process of boarding up its façade, and exclaims: "But everything is closed! This is insane!"

The seagulls are getting anxious, also. One can be heard shrieking: "Agh! Shoot me!"

Teenage boys and girls, or maybe they are older but only look young because they are dressed as cheerleaders, flash pom poms and batons and jump around in a mockery of formation. A gawky guy in the front is calling out gleefully, with a chorus to call back to him:

– I don't wanna work no more!
– This is what we're fighting for!

– So go ahead, let's smash the state!

– Wouldn't that be really great?

Nearing the top of the hill, there is a final turn-off, where the parades end up, and where those ready for confrontation veer toward the fence, and the parliament buildings behind it, where the formal meetings are being held. There is a traffic signal, with two young men swinging from it. A sign reads: "Gauche pour la démocratie" [Turn left for democracy] with an arrow pointing toward the front.

A young woman snaps photos while furiously spinning around in every direction, at this intersection. "Don't forget, it is images that win the war," she explains intensely to her friend.

At the front line, almost all are clad in black, though nearly every flag imaginable, nation-state or otherwise, appears to be represented, held high by someone. And the raging grannies, notably unmasked, dressed in old-fashioned pastel gowns from their ruffled necks to their petticoat ankles, their gray and white hair done neatly in buns and bonnets. A heavy fence stands 12 feet tall, with some people attaching flowers, or ribbons, or paper messages, or love locks, and some climbing up and then being prodded off. Behind the fence, the prodders, a phalynx of riot police waits to lay leaden hands on more protesters, ready to receive word to lob a canister of tear gas, fire a rubber bullet, make way for the water cannon, or whatever other strategy might be next in the protocol.

"Libérez-vous!" [Free yourselves!] the front-line people shout at the police through the fence. "Sautez le mur!" [Jump the wall!]

"Attention!" is called. It comes from the police on the other side. We look up, and a protester has thrown a 750-ml plastic water bottle high into the air. It falls down on the other side. Pop.

Off to one side, a young woman, backpack and heavy boots, asks a nicely dressed older woman, first, if she speaks English, next, if she knows what is happening. "Oui, c'est ma ville," [Yes, this is my city], the older woman replies.

Crack.

Screams go out and a big crowd thunders back down hill, at first frantically, then with many hollering – "Don't run! Just walk!" They are coughing; many are scrambling to the dwindling snowbanks, grabbing fists full of snow to soothe their eyes. The first canisters of tear gas have been lobbed at the front lines of the assembled protesters. The first aid helpers appear, with their squeeze bottles and their cotton cloths to soothe stinging eyes and throats. The retreat ceases. Many can be seen donning their masks, goggles, scarves, and other coverings, and more bodies fill in the gaps heading back toward the front line. Red flags, some with "ya basta" emblazoned on them, flood the streets and provide the next wave.

The same older francophone woman finds the same young woman after this big movement has happened, and asks her, with a shake in her voice: "Ils ont tiré dessus?" [Did they fire at them?]

– FPWP – what does it spell? No rights!

On the roadway, while the path is cleared during a wave of retreat, someone has chalked the message: "We are on the right side of the wall." And someone else has added: "But not for long."

A new scene marching through: a group in formation, wearing black and grey suits with drawings of dollar bills and bar codes sealing their mouths, walking in choreographed synchronicity, and a large marionette pulling up the rear.

A tall blonde man looks at his taller, blonder son, with wild eyes, and rests a hand on his shoulder momentarily, to tell him: "When I called your mother last night, I told her I realize now that I've missed my calling for 30 years! I always knew I was an anarchist, but I never knew how much I would love acting on it!"

Just a few paces away from this pair stands an older man, in glasses, thinking that he perhaps will be diagnosed with throat cancer when his lab results come back. Thinking about his second home, a cottage in the woods, that he has worked hard to secure as a refuge for his private thoughts. Perhaps in the future for his convalescence, maybe his final days. Maybe a place where his children might come to see him one last time. And weighing the likely personal consequences of his activism. Particularly in that after his last protest experience, if he gets arrested again and if an interested party chooses to push forward with a large lawsuit, he might not be able to maintain his mortgage.

There is dancing too, in pockets of land carved out from the hillside, decorated with decades of graffiti, adorned with all manner of propaganda, body painting stations, and with DJs spinning records and Hare Krishnas cooking and serving plates of vegan food. They are a band of spindly change agents moving wildly, kicking up dirt and snow, passing around drugs and bottles of beer. And, as the sun sets, the contrast of this saturnalia with the scene at the top of the city, the mob of black-clad people still hanging on to the front line, some carrying torches, some throwing colourful flares, and each new gas canister launched catching the colour of the flare and making it spread further and last longer.

> – This is what democracy looks like! Chants one group, clapping rhythmically.
> – This is *not* what democracy looks like. Spits an individual in disgust, turning her back on them.

Two people hold a banner up between them that reads: "Inequality is unacceptable." A third person hands out slips of blue paper imprinted with dense text that begins with the question: "How can we decrease social distance in our communities, in our world? This should be the goal of all social policy. Surely we can all agree to this." It continues, half in English, half in Swedish.

A pair of men can be seen crumpling up small blue squares of paper and throwing them down at their feet. One says to his partner: "Equality is unacceptable, without reward and incentive, we lose all the motive force of our movement. We need only to foster greater motivation."

A local man stands within sight of the front line, watching the line of riot police. "Ceci ne vous inquiète pas?" [Doesn't all this worry you?] a visitor asks him, backing up slowly as the riot police advanced.

> "Inquiet? Non," he replies, nonchalantly. "Not as long as the fence stays up."
> "Do you think it will stay up?"
> "Today, yes."

A few members of the media are interviewing groups of those assembled, holding cameras and microphones. Some people scream and then there is a pause at the sound of one television van's window being smashed. Others scribble notes to themselves on small notepads. A few of those who walk by, on their way back to the front line, mutter to them: "Screw the summaries."

An ostensibly indigenous woman has emerged from the underbelly of the river dragon puppet, which has now snaked its way up and down the hill several times. She has a megaphone. She implores, moving in a wave-like motion to guide the river dragon forward: "We will never be able to do it alone because we share this earth, the air we breathe, the water we drink, the garbage we create. It is imperative for our development as human beings to figure out how to work together."

Solidarity. A challenging learning process, to be sure. There is a threesome of men locking arms in the street: one dressed as a priest, one as a rabbi, and one as an imam. The priest can be heard explaining to a few young people, while his cohort nods in agreement:

> You can't be Catholic by yourself, it is instead about seeking community. Was the real miracle that Jesus fed a crowd of 5000 followers with five loaves of barley bread and two fish, by magically expanding the community? Or was it that Jesus brought the people out of their feelings of concern for their individual, personal wellbeing into a feeling of comfort within community by openly declaring that they would all stop and break bread together, and that it was this that caused all the people to reveal and share what they had brought with them with one another? What they had thought they brought for themselves alone, until Jesus made them willing to divulge to others in the group as shared?

The Prime Minister of Canada comes out in front of the riot police, but behind the fence, in order to address the crowd directly and to implore the assembly to disperse. When no one appears to listen to his entreaty, he calls in desperation: "Soyez bons chrétiens!" [Be good Christians!]

## Ordinary justice in the urban commons

This peek into the old city of Quebec, on the occasion of an imaginary, momentous meeting of world leaders and counter-leaders in the service and disservice of progress and world future, gives us a notion of the complexities involved in seeking sustainability and justice. The specificity of this space, at this moment, with this constellation of actors, gives the site a constellation of justice opportunities, a fertile ground for the pursuit of justice because it is a ground already alive with the hopes, indignation, fears and agonies of the feet trampling it at present. At the same time, the pressures and contradictions offer a sense of the constraints on justice-seeking, even in this relatively favourable context of enlivenment. Within such a space and such an understanding, according to Soja (2010: 7), the work of justice-seeking is to promote "more progressive and participatory forms of democratic politics and social activism, and to provide new ideas about how to mobilise and maintain cohesive collations and regional confederations of grassroots social activists." At our particular starting point, we have the city liberated from and broken free of its daily routines, and at the same time locked down on high security alert. We have thousands of people prepared to change their habits, overlapping with thousands of habits that they are not prepared, positioned, or conscious of the commensurate need to change. We see the interactive ripple effects that many such kinds of change could have and are having. Here, the city is clearly not just a set of structures, but an oeuvre as Lefebvre (1991) would have it, a creative work in progress that allows our collective 'right to the city' to take shape, in space. The city, that is, is not simply the stage upon which the work of justice-seeking unfolds; the city is the result of this work, in all its aspirational, active, counteractive, neutral, and ridiculous forms, all of them political (Fournier 2013: 443).

If, with this understanding, we work toward urban justice, we are working to promote "greater control over how the spaces in which we live are socially produced" (Soja 2010: 7). At the same time, we can also recognize that they are more-than-socially produced, with hosts of structural and institutional constraints and tools imposed, different relations to sustainability, perceived differently by different groups. The invisible as much as the visible structures and relationships create opportunities to construct urban public cultures of justice (Amin 2008). Pleading for greater social openness, according to Amin, can only get us so far. Examining the structures that dictate what is normal, what is comfortable, where the risk lies, and how this is internalized and operationalized within the habits and routines of the daily life of urban public spaces, may get us farther.

Ordinary justice can take shape in just such practices, in just such spaces. Ordinary justice can exist within a context of local democracy, as an ethical conception that undergirds local governance as well as social and cultural association (Dewey 1969 [1888]: 240). Part of the challenge of considering the prospect of ordinary justice is to "get rid of the habit of thinking of democracy as something institutional and external and to acquire the habit of treating it as a way of personal life" (Dewey 1988 [1939]: 231), a moral ideal for pursuing justice in ordinary circumstances.

From a pragmatic perspective, formal, delimited and extraordinary forms of justice depend upon a sense of ordinary justice. As such, our approach to seeking justice within the sustainable city begins within the public spaces of the city, the urban commons. We will focus our attention on the notion of the commons, as both a special kind of place and a special kind of social arrangement, and we will examine some of the ways in which the commons takes shape, some of the ways in which action is coordinated in such spaces, and we will look into the differences with regard to the manner of coordination that it permits. The commons, properly conceived, calls forth a kind of action that is particularly suited to democratic progress. We will explore a variety of commons and of communications of inequality and injustice that people perceive to take shape in these spaces. We will hold the commons to be a primordial, essential and critical component of building up an ordinary notion of justice which has the power to hold us captive.

At the same time, we will acknowledge and examine the forces within our contemporary cities that devalue the commons in comparison to other aspects of life, and the work done by these forces to advance regimes of human conduct that ward us away from the commons or from "commoning" the spaces of our cities and our lives. Beauregard and Bounds (2000: 243) recognized that pressures to devalue the commons as a site of justice-action come from at least two directions: "From without, the allure (or the aggression) of other societies, whether real or imagined, threatens. From within, smaller-scale affiliations such as families and over-arching affiliations such as ethnic groups erode the strength of societal commitments." Nevertheless, between the push and pull of these other compulsions, the commons within the city has a shot at leveraging a share of loyalty and responsibility sufficient to build and rebuild a good society. In this way, the urban commons may be the key integrative device of our times, both as a concept that continues to have moral and ideological pull upon many of us, and as a set of structures and institutions and relationships that continue to exist throughout the cities in which many of us live. Work in the commons does not depend upon trust in formal local government. It does not depend upon one's stance on private property or private wealth. It does not depend upon any particular social or institutional positioning. Beauregard and Bounds identify five key clusters of rights and responsibilities that this work does depend upon: (1) safety; (2) tolerance; (3) political engagement; (4) recognition; and (5) freedom. Holding onto these values, the work remains to re-establish the position of the commons, from this basis, as critically important in helping us meet the normative ends of sustainability and justice, in the practice of local democracy.

## What do we know about planning for justice in the sustainable city?

Susan Fainstein has endeavored to keep questions of justice central and core to urban and planning theory. Fainstein has led the charge within an ongoing debate about the nature of the highest and best ends we should pursue when we pursue planning in the public domain. Planning theory has also effectively considered the question of how to incorporate justice in terms of the procedural as opposed to substantive aspect of planning, that is, justice in how planning is done, not what planning does (Healey 1997; Forester 1999; Fainstein 2000; Campbell 2006; Christensen 2015). Campbell (2006: 102) has called for a just planning process based upon "situated ethical judgment," in which "you have to have an understanding of universal values before you can determine the importance of the particular." This position assumes that facts and values "reside in separate ontological spheres discernible through disparate epistemological practices." This distinction between just planning outcomes and just planning process was articulated by Marcuse (2009) as the difference between "justice planning" and "commons planning." Within justice planning, Marcuse argues that planners' ability to tackle the structural causes of unjust and inequitable outcomes in cities is limited. Commons planning, by contrast, opens up the study and confrontation of embedded power relations that determine these unjust outcomes, thus creating space for a process-based deconstruction of how to prevent further injustices from arising in the future, under similar contextual conditions. This distinction between justice as an end versus a means of planning raises important considerations for the subject of this book.

Susan Fainstein (2010) argues, compellingly, for justice to be considered as the highest and best outcome of planning, to be evaluated in terms of universal standards of equity, diversity and democracy. In a hot, crowded, full, past-the-tipping-point, gap-wedged world, this seems to be a most humane and reasonable approach to reduce suffering. Fainstein (ibid.: 8) offers justice "as the governing norm for evaluating urban policy" and specifically articulates a principled position for justice in this way: "Making justice the first principle by which to evaluate urban planning and policy is essential" (ibid.: 10) and "the principal test is whether the outcome of the process is equitable" (ibid.: 13). She demands that just outcomes of planning and policy evaluations amount to meeting standards of equity, diversity, and democracy. After all, as wealthy and clean enclaves of the world enjoy the spoils of globalized capitalism, a growing number of cities are suffering the fallout in terms of pollution, resource degradation, extreme climates, extreme religious commitments, social and health and economic damage. Justice in Fainstein's formulation would provide better emergency care, better long-term treatment, and a phase-out of polluting facilities in these cities, for people who are suffering the most, first. This is a view of justice through planning but not in planning, or as Lake (2016a) offers by way of interpretation: "justice is planning's consequence rather than its core."

This sounds like a straightforward enough principle for planning practice. However, the demands of sustainable development complicate matters considerably. Sustainability expands our definition of justice into the future, demanding that equity with respect to future generations be considered in our justice allocations. Sustainability also extends our definition of equity toward beings not usually considered as within the bounds of those with whom equity is shared: nonhuman species. These two extensions of our community for the purposes of determining equity in a given context essentially make Fainstein's proposed reliance on *ex ante* evaluation for the determination of equity an impossibility. We could not possibly know enough about the needs of future generations, in general, nor about the needs of nonhuman species to permit a general *ex ante* allocation of equitable benefits to both, as well as to those within our purview. We could not possibly devise a reasonable sense of how to allocate justly to the unborn, and to the many nonhuman creatures that share a stake in any given situation.

This is one practical reason why a justice and sustainability approach has to reject Fainstein's Rawlsian proposal for allocating just deserts. There is an ideological reason as well. From a pragmatic, democratic perspective, to attempt to determine just outcomes at the outset of a process, ex ante, gives up on a significant opportunity that what we can arrive at in the course of the process may open up new arrangements and opportunities for justice that we were not able to realize or expect in advance. That we might learn, about the situation, about ourselves, about the future and what it holds in a way that allows us to reconsider the specific meaning of justice in the process of planning. As Lake (2016a) argues: "Viewing justice as the subject of planning makes an explicit consideration of justice a central element within the planning process." Rather than planners seeking justice through planning practice, planning becomes a justice practice. In proceeding in this way, the possibility of emergence is recognized. By emergence, I mean the possibility of an outcome that was completely unexpected before it came to pass.

The ideological reason to proceed in this way is the experimental and evolutionary basis within pragmatist thought. To Dewey, "outcomes are only and already prefigured in the process of their creation" (Lake 2016a). The pragmatic method also rejects what is called subject/object dualism, making the whole task of separating out means and ends futile. Further, the only people suited to carry out the task of evaluating the ends of plans are the same people as that expansive and diverse group that, hopefully, we have engaged in the planning process itself. It makes no sense for standards, in this context, to be set in advance by any group of expert nonparticipants. To separate out these two activities of planning is for planners, on the one hand, to abdicate a commitment to local democracy, and on the other hand, to fail in the challenge of constructing and maintaining a regime of engagement. It is to choose to engage in a kind of talk that is only legislative, not in the kind of talk that is conversational, meaning ordinary, and meaning that which constitutes our social worlds. According to Fraser (2009), talk can be legislative or conversational, and planners have a choice of legislating the meaning of justice or participating in a conversation aimed at formulating justice as a value guiding action in planning. Here, in considering the prospect for an ordinary sense of justice and sustainability in today's cities, we focus on the conversation.

Instead, what Lake (2016a) proposes is that we pursue justice in planning by asking, as a single and indivisible question: "What are we doing and is it just for us to do so?" Pragmatically speaking, this question is ever contingent, situational, and can only seriously be addressed when presented with a set of material circumstances. Justice looks very different under different circumstances. Justice can never be settled once and for all. Justice does not sit "out there" waiting to be achieved. From a pragmatic perspective, then, the first and monumental piece of work in justice-seeking is seeking to understand the specificities of context, conditions, and perspectives which bear upon a situation in need of a plan. Participants will articulate "a variety of justice claims, the sources from which they spring, the types of justifications forwarded in their support, and how these claims are mobilized and deployed to affect planning outcomes in specific settings" (ibid.). To make our social spaces more just, we need to understand the relative positions and perspectives that people and groups are starting with, and to place these in conversation with one another.

For justice to be the guide for planning processes toward desirable moral ends, planning processes need to be attentive to the question of "whether our collective endeavours are producing a material reality in alignment with our moral purposes" (ibid.) – essentially a test of the authenticity of our engagement. Is the engagement we have with others helping us as participants in the social sphere to walk our own talk? Planning practice in this view is a practice of "situated moral inquiry ... aimed at formulating a principle of justice to guide a solution rather than importing a universal principle of justice to apply in an act of final moral judgment" (ibid.).

As a guiding philosophy of social change, pragmatism offers a hand-up in this work as a set of arguments and resources for coping with the shift in time and space horizon from the eternal to the near-term future, and from the global to the contingently local and public. There is great hope in this perspective, in that it offers more than eternal or universal truth, the prospect of a future orientation in which those who put in their oar will be able to see the distance that they travel, together, within our contemporary, material and moral lives. This metaphorical distance becomes the measure of wisdom, which we can construct through this common work of making preferential choices based on an expanding perception of a complex social and physical reality (Dewey 1919: 43). Indeed, this move shifts the notion of wisdom from one of preservation of conformity with established ways and understandings, to one of intelligent and adept adaptation to changing conditions, eschewing the "crutch of dogma" (Dewey 1916: 211). If we abandon the fool's game of entering into a planning process with a discrete, unchangeable objective, we may gain the possibility of genuinely original and unprecedented results of our common work. For all of us who want more for our work in research and reason than a recognition at the end of the day that we have explained current norms as truths, pragmatism offers clear value.

## Expanding the common sense of pragmatism into a critical pragmatism

Invoking pragmatism as a concept essential to the pursuit of justice will strike many as unusual, and some as heretical. In its contemporary political and everyday usages, pragmatic acts could even be considered those that are blind to justice. The critical, more fulsome understanding of pragmatism that we open up in this book reaches deeper into the pragmatic thinking tradition in order to contest the ambivalence about justice that is often identified with a pragmatic perspective.

In contemporary political usage, pragmatists are the characters acting behind the scenes of the politicians to do the unpopular work that, we are told, must get done. They often carry titles such as Treasury Secretary or Minister of National Defence. They make decisions and implement actions that, in and of themselves, would never be supported by a voting public, using the guise that such decisions and actions are pragmatic. That which is pragmatic, in this context, is that which we seek to avoid fully justifying, believing that the results produced will ultimately provide their own justification.

In everyday usage, to say that an act is pragmatic typically means that it is efficient. Pragmatic acts save time and effort, by taking shortcuts of some kind. A pragmatic approach to making dinner might mean using a microwave oven instead of the conventional stovetop, even though the cook may recognize that taking the time to cook on the stove produces a result with better flavour. A pragmatic approach to determining the winner of a writing contest could mean introducing a pre-screening of candidates based on a factor, like age or address, that is quick and mechanical to determine, reducing the amount of time the judges will need to deliberate on the merits of the submissions themselves. Such pragmatic acts can save money as well as time (a lower electricity bill for the cook, lower honoraria to pay the contest judges), and to this extent, they are expedient. Completing a green building with reference to an independent energy efficiency standard, without actually following through with certification of the building based upon compliance with this standard, is a common approach taken by builders who see themselves as acting 'pragmatically.' Rather than "certified green," the building that results is called "certified green equivalent." This approach saves the time and money that would be spent on the certification process itself, while ostensibly (although perhaps not verifiably) achieving a similar result in terms of green qualities.

Pragmatic acts, viewed in this way, pack a punch. They don't mince words. They don't suffer fools. They don't take up more time or space than they're worth. They make do with what is at hand. They satisfice. They may seem reckless, but they help us get by; they may neglect the recipe, but when they have lemons, they find a way to make lemonade. They are, to put it in another lexicon, entirely consistent with current thinking on resilience. This is a parallel we will make use of in Chapter 6.

And, in another way, taking a pragmatic perspective is sometimes considered optimistic to a point that is unhelpful in terms of confronting injustice. Pragmatism entails a sort of doggedness of optimism, an optimism that is beyond any justification by instrumental success. A pragmatic attitude toward optimism in action, toward the possibility of progress in the democratic sense, is about as close to a foundational principle as this anti-foundationalist philosophical tradition gets. This is the way in which pragmatism will be drawn upon as a stance of empowerment in Chapter 5. Because, regardless of instrumental success in specific outcomes of an act, there must remain the prospect of further learning, further engagement, and further construction and expansion of the material of democratic community building. And from this conviction, an undeniable sense of empowerment flows.

This notion of what pragmatism means is consistent with the concept of pragmatism employed in this book. But, when extended into a more fulsome consideration of the social and material conditions of our urban lives, public interest, and the aspiration for democratic community, and the play of knowledge, learning and perspective-taking, pragmatic acts become radical acts. One of the founders of pragmatic philosophy, William James, articulated this radical shift when he discussed the pragmatic notion of truth as the "cash value" of an idea.[1] He did not have entrepreneurialism or marketing on his mind when he said that what is true could be thought of as the cash value of an idea. What he meant was that societies could do a better job in seeking truth and justice if, rather than basing judgments of right and wrong on abstract principles, we instead base this determination on our own experience of what does and does not have value. A pragmatic approach to urban sustainability and justice, then, entails a search for the cash value of the city for urbanites committed to sustainability and justice, in terms of the interplay of today's cities, and particularly the collective experiences that we have in the urban public realm.

At the same time, in common usage, recognized pragmatic values of efficiency and expediency come at the expense of equity and thoroughness. What happens to determinations of quality, for example, under pragmatic specifications? Can there be no value ascribed to that which can only flow from spending more time and effort than absolutely pragmatically necessary? Can there be no recognition of the value of that which we remain unable or unwilling to monetize or render substitutable according to "cash value"? Pragmatic actions, commonly considered, are tone-deaf to the qualities of different voices that might be raised, or the silence of voices not raised, or the ways in which voices may play harmoniously together in particular arrangements, but clash in others. They are blind to the differences that different processes might make to the experiential quality of the action. They oversimplify the complexities among and between individuals and their specific histories and contexts. This is a real critique of the philosophy of pragmatism, too. It is one that is aptly taken up by the pragmatic sociology of engagements, and which we will draw upon here in order to capture the importance of the different qualities of justifications and engagements and the bearing these have on the prospects for sustainability and justice.

Taking this commitment to the philosophy of pragmatism one step further, research making use of the pragmatic sociology of engagements examines this very process of knowledge generation, the process of forming these principles in particular situations. Dewey held that there was no real difference between the practical skills employed by scientists in experimentation and the testing of new knowledge and the skills needed for the practice of democracy as inclusive social collaboration and experimentation in order to resolve social problems (Lake 2016a, 2016b). The methodological approach of the pragmatic sociology of engagements offers some specific tools to help make this contextual leap.

The pragmatic sociology of critique and engagement begins from a recognition of a plurality of layers of understanding within the socio-political landscape of liberal democracy. Within these layers, people confront requirements from their peers to give reasons for their different understandings and actions. They are forced to somehow justify the ways in which they fault the conditions in which they live, and when this has a community-building effect, this justification is offered in the form of a kind of engagement in the public sphere. In the public sphere, among members of one or more social group, the argument being voiced is determined to be either legitimate or not. Further repetition of the justification and perhaps judgment of the worth of the person who first offered it will be made on this basis.

Communities place boundaries on the kinds of justifications that can be offered legitimately in various situations and conditions. For example, where a debate focuses on a contentious issue of caring for the poor, a 'good' justification in most cases cannot merely appeal to blatant private interests of the well-off, but must respond to an established general interest in common social value. Of course, a range of legitimate justifications can still exist, and these will continue to be disputed in democratic society. Distinct from a Habermasian communicative action approach, the pragmatic approach does not impose a single ideal set of conditions for communicative rationality, but attends to the reality of diverse social and political settings, and the material conditions that actors rely on when they form opinions and arguments. Moreover, a pragmatic approach recognizes that actors engage in a variety of capacities in different issues and debates in the public sphere, and may act in a different capacity each time, depending on what community membership or perspective takes precedence for them at that point. In essence, there is more to understanding public debates, disagreements, and the engaged work of community building than moving from the specific to the general, from the individual to the collective, or from the emotional to the scientific (Boltanski and Chiapello 2005 [1999]; Boltanski and Thévenot 2006).

Different from the conventional understanding of "engagement" in community development and social science circles, typically evoking a specific ideal of participatory democracy by empowered and self-actualizing individuals, "engagement" here is understood as a more encompassing set of activities. Still thoroughly integral to the attainment of conditions of democracy, engagement in the pragmatic view "indicates a relation to yourself through the environment, in time, and not only to the present situation … engaging with an appropriate environment is a condition for enacting a certain beneficial capacity" (Blokker and Brighenti 2011: 12). That is, effective engagement is a relationship between actors and environments, and understanding that engagement demands attention to both. Both actors and environments are constrained in important ways and cannot be persuaded or compelled by theoretical argument about the benefits of democracy to pursue any given course of engagement. These constraints put different arguments into different regimes of engagement – not boxes that cannot be escaped but "adherences" in every form of engagement, from the most intimate to the most functional (ibid.: 13).

The ways in which people express their disagreements can take the form of different "regimes of engagement." Thévenot makes the distinction among three such regimes in order to open up reason to hope for change in established routines and practices even in the face of hardened market-driven conditions. The first regime is that which we expect in academic discourse: the rational interplay of ideas among educated people, engaging as equals, able to identify and seek common understanding with an interlocutor. This regime demands Habermasian ideal speech conditions or what Dewey anticipated as a society-wide community of inquirers in his notion of the Great Community. The second regime is the one that is currently dominated by a neoliberal understanding of value and worth. In this regime, individualistic, profit-making 'idea work' generates value more than other kinds of activities. This is the assumption encountered in many forms of policy and public debate. It is, of course, a regime that is oppressive toward individuals unable to compete on these terms or express their concerns in terms of neoliberal value, which is the case for most community wellbeing concerns. The third regime is the regime of family and emotion, where what counts is what can be shown to be felt most deeply by people we care about and identify with. It is often an irrational regime, but in communities full of personal attachments, it is no less a communicative reality than the other two regimes.

Each of these distinct regimes of engagement involves communication or an attempt to make personal concerns common within a group, which is required "to alleviate the tension of living both together and in person" (Blokker and Brighenti 2011: 8). Each involves the composition of a plurality of voices and constructions of the common good, common places, common sites of meaning. And each of the three regimes has its own specific advantages and disadvantages where community building objectives are concerned. The distinction among these regimes of engagement serves the important function of revealing the shortcomings of modern constructions of individuals, debates, and publics in understanding the workings of justice-seeking in the city. Debates are differentially invested with different sets of analytical tools, and if we hope to keep public debate within the realm of the constructive, let alone socially transformative, and avoid settling into the regressive and reactionary, we need to remain aware of the different forms of engagement. We are not all prepared to generalize, to think of ourselves as capable self-maximizing individuals (even in some idealized way), yet we all have the potential to be political actors. People disagree, and it is important to understand the modes and contexts of their disagreements in order to advance objectives of increased engagement in the public sphere, and possibly to arrive at transformative alternatives. In so understanding, we gain the tools to register the value of protecting and reinforcing human and social diversity within our ideals of the Great Community.

Because global dynamics contextualize and help to determine local circumstances, even as the global is constructed via concrete material actions, there is room for the application of certain *longue durée* sociological classifications of justifications as per a pragmatic sociology of engagement, in order to advance our understanding of what is being proposed as justice, why, and perhaps for how much longer. This is a sound pragmatic approach to research and action in the common interest, because it reflects how we behave every day in the urban environment: we negotiate extreme technological complexity, and market dynamics, a host of social norms and expectations, and other diverse human beings. At the same time, we often seek to collaborate in so doing, and this collaboration is choreographed and improvised behaviour that we have to learn, reinforce, and change, sometimes quickly. Within the resultant scene of human clash of social purpose and aspirations we have the opportunity to experience:

> the significant record of the efforts of men [*sic*] to formulate the things of experience to which they are most deeply and passionately attached. Instead of impersonal and purely speculative endeavors to contemplate as remote beholders of the nature of absolute things-in-themselves, we have a living picture of the choice of thoughtful men [*sic*] about what they would have life to be, and to what ends they would ... shape their intelligent activities.
>
> (Dewey 2004 [1920]: 15)

The thought exercise with which we began this chapter, of sustainability and justice-seekers in the streets of Quebec City in a state of exception, provides layers of understandings of justice and sustainability. There is the understanding of justice and sustainability that the design of a fortified city intends in the very first instance, sited and constructed by the French colonizers to exert power and control over their new world, and to protect themselves and their interests against other invaders and the existing inhabitants. Then there is the understanding of justice and sustainability that leaders of democratic nation states evince, when they elect to convene a global meeting in order to plan for and address long-term and global matters that none of them, acting alone, has the political capital to tackle. There is the different understanding of justice that a civil society occupation embodies, through accompanying and presenting itself in counterbalance to such a formal, multilateral negotiation. Finally, as I hope to have demonstrated by pointing out a sample of the multiplicity of perspectives operating on the ground, there are the very many different varieties of justice and sustainability understood and enacted by those who are sufficiently engaged members of the political community to physically attend the civil society event. The state of exception for our thought experiment helps to bring political focus and importance to the perspectives, questions and actions suited to the public sphere of political engagement, that normally compete with and are crowded out by other kinds of demands that daily projects impose on people.

If we had, in our thought experiment, "jumped the fence" to the formal side of the Forum, we would have encountered yet more versions of justice and sustainability. We did not because we know enough about how formal politics operates, in such multilateral and international meetings. We know less about what happens to notions of justice and sustainability when people meet in urban streets and public spaces, the places where ordinary public life and a common interest can take shape. (And the places, it must be remembered, where pro-nouncements and decisions by elected leaders must be implemented – or fail.) In such places and interactions, advancing justice and sustainability offers more choices than black, white, or green.

This book does not argue that one or other version of justice is correct, nor that one or the other is likely to lead us to a just society that values sustain-ability and justice both within and beyond the human species and within and beyond present-day political sights. What this book intends, rather, is to argue for an approach to understanding and seeking justice that comes from a prag-matic commitment to multiplicity. By starting from a basis of recognizing pluralism in our cities, and the importance of this aspect of our contemporary urban condition to embarking on a project to address justice and sustainability in the public interest, we can devise useful strategies for meeting other people in the public sphere. It is these meeting points, with all their contingencies and conditions, that demand our greater attention, in the interest of justice and sus-tainability. This is a frighteningly risky position to be in, full of traps and pitfalls that critical urban theorists have been trying to warn us about, as we witness the rise of a new celebration of all things urban, in our time. This book does not discount this risk. At the same time, it seeks pragmatic reasons to continue the work of trying to engage in this way, and even plausible reasons to notice particular promise for this kind of engagement in today's cities.

## Note

1  "Grant an idea or belief to be true ... what concrete difference will its being true make in any one's actual life? What experiences [may] be different from those which would obtain if the belief were false? How will the truth be realized? What, in short, is the truth's cash-value in experiential terms?" (James 1977 [1907]: 311).

## References

Amin, Ash. "Collective Culture and Urban Public Space." *City* 12(1) (2008): 5–24.
Beauregard, Robert and Anna Bounds. "Urban Citizenship." In *Democracy, Citizenship and the Global City*, edited by Engin Isin. New York: Routledge, 2000, pp. 243–256.
Blokker, Paul and Andrea Brighenti. "An Interview with Laurent Thévenot: On Engagement, Critique, Commonality, and Power." *European Journal of Social Theory* 14(3) (2011): 383–400.
Boltanski, Luc and Eve Chiapello. *The New Spirit of Capitalism.* London: Verso, 2005 [1999].

Boltanski, Luc and Laurent Thévenot. *On Justification: Economies of Worth*. Princeton, NJ: Princeton University Press, 2006 [1991].

Campbell, Heather. "Just Planning: The Art of Situated Ethical Judgment." *Journal of Planning Education and Research* 26(1) (2006): 92–106.

Campbell, Scott. "Green Cities, Growing Cities, Just Cities? Urban Planning and the Contradictions of Sustainable Development." *Journal of the American Planning Association* 62(3) (1996): 296–312.

Christensen, Karen. "Both Process and Outcome are Essential to Planning." *Journal of Planning Education and Research* 35 (2015): 188–198.

Dewey, John. *Democracy and Education*. New York: Macmillan, 1916.

Dewey, John. "Philosophy and Democracy." *University of California Chronicle* 21 (1919): 39–54.

Dewey, John. "Creative Democracy – The Task Before Us." Reprinted in *The Later Works of John Dewey, 1925–1953*, edited by JoAnn Boydston, vol. 14, Carbondale, IL: University of Illinois Press, 1988 [1939], pp. 224–230.

Dewey, John. "The Ethics of Democracy." Reprinted in *The Early Works of John Dewey, 1882–1898*, edited by JoAnn Boydston, vol. 1. Carbondale, IL: University of Illinois Press, 1969 [1888], pp. 227–249.

Dewey, John. *Reconstruction in Philosophy*. Mineola, NY: Dover Press, 2004 [1920].

Fainstein, Susan. "New Directions in Planning Theory." *Urban Affairs Review* 35 (2000): 451–478.

Fainstein, Susan. *The Just City*. Ithaca, NY: Cornell University Press, 2010.

Forester, John. *The Deliberative Practitioner: Encouraging Participatory Planning Processes*. Cambridge, MA: MIT Press, 1999.

Fournier, Valerie. "Commoning: On the Social Organisation of the Commons." *Management* 16(4) (2013): 433–453.

Fraser, Nancy. *Scales of Justice: Reimagining Political Space in a Globalizing World*. New York: Columbia University Press, 2009.

Healey, Patsy. *Collaborative Planning: Shaping Publics in Fragmented Societies*. Vancouver: University of British Columbia Press, 1997.

James, William. "Pragmatism and Radical Empiricism." In *The Writings of William James*, edited by J.J. McDermott. Chicago: University of Chicago Press, 1977 [1907], pp. 311–317.

Lake, Robert W. "Justice as Subject and Object of Planning." *International Journal of Urban and Regional Research* (2016a). In press.

Lake, Robert W. "On Poetry, Pragmatism and the Urban Possibility of Creative Democracy." *Urban Geography* (2016b). In press.

Lefebvre, Henri. *The Production of Space*. Oxford: Blackwell, 1991.

Marcuse, Peter. "From Justice Planning to Commons Planning." In *Searching for the Just City: Debates in Urban Theory and Practice*, edited by Peter Marcuse, *et al.* New York: Routledge, 2009, pp. 91–102.

Soja, Edward. *Seeking Spatial Justice*. Minneapolis, MN: University of Minnesota Press, 2010.

# 2   Sustainability as a slippery and a sticky concept

Paradoxical though it may sound, sustainability is both a slippery and a sticky concept. It is slippery in the sense that it is notoriously difficult to define; embraced with equal fervour by those who would never agree on the first step towards its attainment. Even among those groups and individuals actively pursuing it, it seems the more they do, the less able they are to claim victory. At the same time, sustainability is a sticky concept in the sense that, despite a quarter century of prognostications of its terminally diseased condition, as a passing ideal, an overly intellectualized pairing of poverty alleviation and environmental goals, a bourgeois fantasy that failed to connect with the material needs of the masses, sustainability has persisted and carries cash value as an organizing frame for policy, planning, organizational management, consumer behaviour, and on and on. It has a demonstrated ability to organize urban policy, planning, and institutional design in new and useful ways. This chapter reviews the many factors at play in this slippery and sticky dynamic and then constructs an argument about how a pragmatic approach offers a notion of sustainability with justice value. It considers the prospect of constructing a justice theory of sustainability building upon pragmatic philosophy and upon the critical pragmatic sociology of engagement.

Sustainability means green: there exists within the concept of sustainability an appeal to the value of nonhuman nature, long neglected in modern human affairs. This has one set of meanings in rural and wilderness areas, areas which are sparsely populated by people. It has an entirely different meaning in urban areas, where the treatment of nature as a matter of political importance is more unexpected. The sustainable city values the best parts of what have traditionally been considered a natural or rural lifestyle – clean air and water, fresh food, daily connections to local context and territory, plenty of time to relax and enjoy the simple pleasures of life – with none of the social and economic limitations also traditionally associated with rural life.

In its slippery sense, sustainability represents conceptual slippage away from pure 'green' and incorporating the needs of a nonhuman environment, to something more aspirational, more integrative, less strictly ecological (Holden 2006). Reed (2007), for one, has suggested that 'green' implies the ideal of doing less or no harm within a design context, while sustainable design reaches toward a capacity to improve, restore, and sustain the health of the social and ecological context over time. Larrick (1997) and, more recently, Hopwood et al. (2005), Mang and Reed (2012) and others have taken a broader perspective. They have situated the concept of sustainability as a dynamic integrator of human and earthly development objectives, and of contributions from both green and regenerative approaches to design. This vision of sustainability demands unprecedented levels of integration of policy components and goals, leading to some complaints that its ambitions to holism over-reach its ability to deliver synergistic value at a ground level (Holden 2012).

Key to the slippery and sticky status of sustainability are its interpretation and use at local and urban scales. The juxtaposition of 'urban' and 'sustainability' has not always been obvious. In the early 1990s, the time of the World Commission on Environment and Development and the first Rio Earth Summit, the combination of sustainability and cities was considered anathema by many. Rees (1992: 125), in presenting the ecological footprint concept, put the relationship in starkest terms. Rees offered the concept of ecological footprint as a standard measure of the amount of ecosystem capacity appropriated by a person, or grouping of people like a city, as a new means of accounting for the burden of human lifestyles on natural carrying capacity. His bleak assessment was that: "However brilliant its economic star, every city is an ecological black hole drawing on the material resources and productivity of a vast and scattered hinterland many times the size of the city itself."

In the intervening quarter-century, something drastic has happened to the placement of the nucleus for sustainability work. Without any evidence being presented to negate the basic validity of Rees's understanding of the city–environment relationship, the political presentation of what cities mean to the prospect of sustainability has nevertheless shifted completely. Where the logic within the sustainability movement was once that "small is beautiful" (Schumacher 1973), big is now in. Now, urban political leadership, local initiatives, and urban experiments are considered absolutely central to any sustainability agenda; by some, cities are considered to be the exclusive domain of effective sustainability action. The slippery and sticky qualities of sustainability work well within the shift of our understanding of political power in contemporary times. At one time, power was considered to amass and organize hierarchically, with nation states on the top and cities and local governments below. This view privileged national and international processes in support of sustainability. Since the 1990s, however, increasing attention has been paid to the ways in which power can also move and accrue laterally, with "governance capacity" (Healey 1998), "mobilizing power" (Healey 2004), and "collective impact" (Kania and Kramer 2011) more likely to be achieved at the local level, where we all live. This shift has privileged work at the local scale, at a time when urban populations have been increasing. This shift also emerges following a period of deep frustration among sustainability advocates and leaders for unified action through national and higher-order processes. Governance power in cities, appealingly, can be flexible, and can be exercised across traditional policy sectors, actor categories, and legal frameworks. City governments can create new integrative policies with fewer hurdles than national governments, and within their more limited jurisdictions and geographical boundaries, retain a glimmer of hope in a unified goal and target for concerted action.

This glimmer, however, may represent just another dimension of sustainability's nature – a reflection off some source more than an actual light. Jaccard et al. (1997), referring explicitly to policy and action to combat climate change, make a compelling case for the blindness that an adherence to urban strategies can bring, since even the most effective action at the urban scale can be completely overwhelmed and negated by status quo inaction at higher orders of government. If cities have, in embracing a slippery and sticky sustainability agenda, shed their reputation as "ecological black holes," they may well have assumed another astronomical identity as "ecological supernovae" in the sense that aspiring sustainable cities may cast a bright green light over the eyes of all those within their force field, blinding them to the ongoing political patterns of parasitism and destruction beyond the urban containment boundary, as well as to the injustices wrought by decisions in the name of sustainability within the city.

Is urban sustainability, then, simply the next best hope for change advocates and local politicians seeking to heal their wounds and frustrations from failures incurred on the national or global sustainability stage, a few decades ago? Or is there, instead, any real reason to expect that an urban focus can make a difference to sustainability outcomes? The effects of sustainability work can now be observed on the ground, in the lived spaces of today's cities. At least, this is the case in the thousand-plus cities that are explicitly using the language of sustainability in planning, development, and governance. Wilson (2015: 2) refers to the interest among such cities in sustainability as a "tidal wave" and to the work of sustainability as presently shifted "from a dominant movement to a near orthodoxy." In this sense, embedding sustainability within a city assumes the look and feel of vibrancy, health, sociability, and vitality. Urban sustainability takes a wide but not infinite range of material forms that include cycling and walking infrastructure, smart growth-style neighbourhood design with smaller yards and larger front porches and stoops than typical suburban designs, thinner streets with more people lingering on them, local businesses and food markets, waste reduction and waste management strategies, and public art that calls for carbon awareness, among other things. The ambiguity – the slipperiness – sits between these specific and mundane material elements of sustainability in cities, and the ambitious policies that use highly-aspirational language like post-carbon, de-growth, zero-waste, inter-generational, one-planet, and other monikers of global virtue.

Politically speaking, this positioning of sustainability has proven extremely sticky, in the sense of holding demonstrated value and attracting a large and growing set of political leaders to use the language and strategies of urban sustainability. The orthodoxy of mixed-use and medium- to high-density urban development, for example, views parks and green spaces, and urban food systems, as indisputable "goods" for cities, despite the fact that most if not all contemporary cities are dominated by strikingly different land use patterns. This gap between desire and reality is a testament to the sticky political nature of sustainability. Key also to this stickiness is the way in which the orthodoxy of sustainable cities casts a shadow over the offshoots and casualties of sustainable city strategies, some of which may be accidental, but some of which may also be systematic: such as a middle-class bias, blindness to many kinds of social difference, gentrification, and a secular Western vision of self-fulfilment. Significantly, urban sustainability success does not appear to depend upon demonstrating a divergent economic base from the fossil fuel-driven economy which lies, in virtually all accounts, at the root of the sustainability crisis globally. The stickiness of sustainability comes as a result of its political success in demonstrating vision and concrete, design appeal, and simultaneously hiding the inconsistencies, blind spots, and gaps within the expression of the idea.

In a review of the research and practice concerning the role of cities in a sustainability transition, Bulkeley (2010) notes that there has so far been a tendency to perceive the political contexts in which climate policies should be realized as relatively coherent and favourable to change. She argues that there is a need to develop a more critical approach and a better understanding of the political and economic contexts in which negotiations on climate and urban development take place. Researchers are right to continue to question whether there is meaning in the moves toward a better understanding of the fit between sustainability policy, climate change-informed urban policy and a more traditional urban policy framing, which favours traditional, and unsustainable, notions of social and economic development (Finn and McCormick 2011). Sustained theoretical attention to the role and status of sustainability as a social theory *per se* is needed in order to push our understanding of sustainability further. Nor is this a matter of solely theoretical importance; it is crucial to developing an understanding of the extent of political work possible under the banner of urban sustainability.

Critical social scientists have been drawn to this simultaneously slippery and sticky sense of sustainability in order to point out the way in which the concept serves, perhaps in a socially and ecologically sophisticated way, perhaps with new political power, as the same kind of cover for the interests of capital accumulation that has been invented many times before. Sustainability sticks because it makes the process of capital accumulation seem worthwhile and socially useful again, after the previous crises of the industrial and post-industrial city. Sustainability is held aloft for its holism, its ability to integrate environmental, social and economic concerns, because the slippery nature of this concept allows it to serve as a 'fix' for the contradictions of capital (While et al. 2004). This opens up new channels into which capital may flow. Critical social scientists have examined how the 'sustainability fix' or, in specifically urban design and development terms, the "smart growth machine" (Gearin 2004) represents a repackaging of the cyclical, continuous work of capital in colonizing human social life (Swyngedouw 2007). In an earlier phase of capitalist urban development, the crisis of industrial economic development was 'fixed' by the post-industrial, service, technology, and tourism economy; so sustainability is the new fix for the next crisis. Or, just as the growth machine of political and business leaders works for them together toward a common goal of maintaining growth, so the sustainability planning concept of 'smart growth' fixes the work of this growth machine toward particular patterns and forms more aligned with the ostensible look and feel of urban sustainability, less with that of suburban sprawl. That is, the overarching and slippery thrust of sustainable development is one of political economy: "sustainable development is itself interpreted as part of the search for a spatio-institutional fix to safeguard growth trajectories in the wake of industrial capitalism's long downturn, the global 'ecological crisis' and the rise of popular environmentalism" (While et al. 2004: 551).

The thrust of this and related social science scholarship offers healthy skepticism about the promises made in the name of urban sustainability from a political economy perspective. This view offers critiques of neoliberalism and governmentality in sustainability's clothing (While et al. 2004; Swyngedouw 2007; Rydin 2012). Aside from these perspectives, the primary debate concerns whether sustainability as a planning frame detracts from or enhances the regulatory, democratic, and environmental goals of planning (Campbell 1996; Cowell and Owens 2006; Gunder 2006). The urban sustainability literature itself engages in interminable definitional debates and offers a range of more and less sophisticated tools, metrics, and case studies of cities and sustainability transition processes (Pezzoli 1997; Fitzgerald 2010; Birch and Wachter 2011; Dale et al. 2012; Lorr 2012). Many of these are somewhat slippery, but often sticky, celebratory studies that forge ahead with the work of urban sustainability. Shrugging off the slippery and sticky nature of the concept as something that can never be completely grasped, nor completely cast off, few bridges have been built. Some scholars go so far as to embrace the ambiguity of the term sustainability as more constructive than it is delusional, confounding or politically permissive (Robinson 2004). In this view, sustainability can usefully be thought of as the emergent property of a conversation, or dispute, about desired futures that is informed by some understanding of the ecological, social, and economic consequences of different courses of action (Robinson 2003, 2004; Robinson and Tansey 2006).

Of course, not everyone who uses the term sustainability engages it as an essentially contested, potentially transformative concept – that is, as a concept that cannot be defined scientifically or absolutely but that has deep cross-cultural meanings and can find different expressions in different times and places. But some do. The daily lives of many urbanites as well as deeper sustained analyses offer abundant evidence that "social justice and environmental sustainability are not always compatible objectives" (Dobson 2003: 83). It is patently obvious to those with the will to see that seeking environmental sustainability can ignore the poor and perpetuate injustice through neglect; and can have negative impacts on the poor, when property values and the introduction of middle-class values to formerly neglected areas make poor people unwelcome where they once lived (and suffered) (Curran and Hamilton 2012). In the inverse order, there are also no guarantees that policies and movements for social justice will result in greater environmental sustainability. Dividing up and redistributing environmental 'bads' such as landfill sites, waste transfer stations, or dirty industrial locations between wealthy and poor neighbourhoods may result in a fairer outcome, but there is little reason to expect that this will result in less waste or pollution overall. "The answer," according to Dobson (2003: 91), "to the question of whether justice is functional for sustainability is: it depends. It depends on the type of justice we are talking about and on the conception of sustainability at stake."

We may think of justice as limited to human beings and, further, to those currently alive, and within an understanding of their current capacities and aptitudes for understanding their own wellbeing, and, even with privileged consideration, given to those who have suffered disproportionate injustice historically. If this group or these groups are the basis of our consideration for the just outcomes of our actions, then seeking social justice within sustainability efforts puts a limit on the relative value that can be sought on strictly environmental terms, without compromising social goals. If, for example, saving trees sacrifices human livelihoods or outlawing factory farming prices some people out of meat to feed their family, social and environmental goals within that particular framing exist in a trade-off relationship. The co-existence of social justice and sustainability goals in a case like this even puts a limit on the value that could be given to social terms that privilege a sustainable future, at the cost of a sacrifice right now. This could be the case, for example, if we were to imagine a ban on private automobiles and parking spaces in order to preserve a right to remain for poor people in a neighbourhood being redeveloped for sustainability. Once unborn people with lifestyles that are currently unimaginable, nonhuman animals and the nonhuman natural world are excluded from the community of justice recipients, there is neither 'justice to the environment' nor 'justice for sustainability'; although there may still be increased justice within a certain community of human beings. In other words, the demand from some environmental justice advocates to bring central concerns of underprivileged people back into environmental politics might marginalize or even preclude some kinds of environmental and sustainability outcomes. In this sense, just as sustainability efforts have been criticized for creating unjust results, the implementation of social justice principles might not necessarily lead to an increase in environmental sustainability.

Given the prevalence and problematic nature of urban sustainability as a dominant frame within politics, society and culture, the paucity of scholarly effort to find a middle path for sustainability that improves the prospects of justice is somewhat surprising. Despite scores of references to a multitude of different notions of sustainability across the gamut of social sciences, rarely has the concept been tapped as a deeper cultural signifier, a concept of an order of magnitude fit to hold a coherent set of understandings of how to behave well, in relation to one another and to our world. It is not enough to consider how aspects and offerings of these different approaches could be brought together in the spirit of interdisciplinarity to result in a productive urban sustainability research agenda that also offers a means for research to influence policy and practice positively. Evans and Marvin (2006) document the trajectory of inter-disciplinary efforts by research councils in the UK in relation to the sustainable city, and provide a compelling argument for why researchers should not count on the fruits of interdisciplinarity.

An alternative to interdisciplinary marriage of separate fields is the striking up of new fields at intersection points or on the margins. Following on the work of religious, community, and labour leaders in the Southern United States, the study of environmental justice examines the political processes and implications of the distribution of environmental 'goods' and 'bads' within environmental decision-making (Agyeman et al. 2002). Agyeman et al. (2003) then launched the project of conceptualizing a just sustainability by foregrounding issues such as human rights, equality in access to resources and facilities, and the production of economic opportunities and decent quality of life for all into sustainability policy and practice discourse. Other constructive debates for the purposes of understanding justice within sustainability are forays into environmental politics and urban envir-onmental justice, from whence emerge concepts like ecological modernization (Hajer 1995; Dryzek 2013), the post-ecological condition (Blühdorn 2000; Blühdorn and Welsh 2007), and green citizenship (Eckersley 1992).

The pragmatic approach to justice and sustainability begins with this need for continuous and ongoing questioning, from as wide a range of perspectives as can be advanced and defended in public. By some accounts, sustainability represents an "essentially contested concept," an idea so big and so central to our sense of ourselves as human beings in society that we can never expect to agree on a single specific meaning (Jacobs 2006; Connelly 2007; Ehrenfeld 2008; 2009). By other accounts, sustainability is, more modestly, a "primitive concept," "introduced where the need is felt without explanation or examination of its character or credentials" (Becker 1960: 32). As a primitive concept, sustainability resists specific definition because the role it plays in discourse is tied in sig-nificant ways to its ambiguity and common-sense vagueness. Some, like Bruno Latour (2008), argue that sustainability is an attempt within modernist ideals to set our sights at what is beyond our grasp. Perhaps urban sustainability is, in this view, another type of glue to the stickiness of the modernist ideal in designing and conceptualizing cities, yet still too slippery to be placed in real urban contexts.

In any case, a pragmatic understanding of sustainability demands a process-based approach. A process-based approach acknowledges the inherently normative and political nature of sustainability, the need for examination and integration of different perspectives, and the recognition that sustainability is a process, not an end state. Sustainability is emergent, based on provisional understandings and decisions about the nature of the world, pieced together in the public sphere. By this view, sustainability must be constructed through an essentially social process whereby scientific and other "expert" information is combined with the values, preferences, and beliefs of affected communities, to give rise to an emergent, "co-produced" understanding of possibilities and preferred outcomes (Robinson 2004: 381).

From a pragmatic perspective, this deliberative conversation is essential, on an ongoing basis; but it is insufficient. Also needed is a better sense of methods and boundaries, because urban sustainability is an inherently normative and political concept, because deliberation is a social process that has to be able to operate in contexts of extreme diversity, and because within this diversity of perspectives, understandings and biases lies the diversity of prospects for integration and alternatives.

I propose here that sustainability should be pursued in the city as one dispute about norms of justice, as an evolving model for ordering the world of goods and bads. The question is whether we, any of us who care enough to engage in the conversation, can see evidence of a commonly accessible root of sustainability as an orientation to a good life in a public sense. We will also return later to the question of what this delimited conversation has to do with the city, specifically; and conversely, what the city might have to inject into this conversation in its own right. We continue with a brief introduction to the basis for thinking of sustainability as a pragmatic vision.

## Pragmatism and sustainability

If we accept the idea of sustainability as either an essentially contested or a primitive concept, sufficiently deep and resonant that it should compel our public attention, then we accept that sustainability is a concept we can better understand by understanding its differences. This transformative expectation of sustainability can be found in practical use. Examples include One Planet Living, a development framework that uses ecological and carbon footprinting to embed an understanding of living within limits into urban developments on the ground (Bioregional 2016), degrowth economics (D'Alisa et al. 2014), and intergenerational planning (Miller and Siggins 2003). This is the understanding of sustainability that can make a difference to our societies, and to their justice. This is the understanding that we will employ in this book.

The philosophical movement of American pragmatism began with the Metaphysical Club in Cambridge in the 1870s, was announced as a new philosophical perspective by Charles Peirce (1839–1914) and William James (1842–1910), and was expanded considerably by John Dewey's (1859–1952) Chicago School of pragmatism, which was influenced in turn by Jane Addams's (1860–1935) social and peace activism.[1] In the early part of the twentieth century, pragmatism was an active philosophy of social conscience and reform (West 1989; Addams 2002). It subsequently fell out of favour in the post-war period, but has more recently re-emerged, particularly in planning theory (Friedmann 1987; Blanco 1994; Hoch 2002; Verma and Shin 2004; Holden 2008; Lake 2016) and public administration literatures (Garrison 2000; Shields 2003; Miller 2004). Pragmatism is a rich and varied intellectual tradition. At core, pragmatists share a commitment to the essential uncertainty of life, to relationships rather than individuals as irreducible units, and to the notion of praxis (the notion that knowledge must be acted out in order to be held). To a pragmatist, "beliefs are really rules for action," (James 1977c [1907]: 377) such that the significance of a thought's meaning is the conduct that thought is likely to produce. Thoughts are only useful to the extent that they can be differentiated by possible differences in practice.

The idea that holds the most revolutionary implications for consideration of urban sustainability and justice is the pragmatist concept of 'truth'. As a public philosophy that brings inquiry out of the realm of a private search for truth, pragmatism searches for thoughts able to guide social action in morally relevant ways and to answer: "How does this help?" What is true depends on context as much as it depends on correspondence with facts: "Truth is not only a matter of logical agreement between different elements of knowledge; it is also the psychological agreement between messenger, the message, and its receiver" (Verma 1996: 6). The implication is that more attention needs to be paid to the consequences of adopting a particular version of the truth, rather than its correspondence with some "for all time" validity.

This is a challenging and frightening idea even to consider, let alone adopt. As an anti-foundationalist posture, this relationship to truth seems to set our sense of justice and sustainability adrift in a sea of relativism and incommensurable values. Countering this, pragmatist Hans Joas considered it a trick of classical metaphysics in philosophy that most of us feel a need for more solid metaphysical foundations to our notions of truth, in order to feel less vulnerable, less overloaded with conflict and uncertainty. This trick holds most of us back from considering the prospect of improving our own wellbeing through embracing a more radical, pragmatic, notion of truth. Joas considered it a secondary trick that the increasing ability of human society to dominate the planet and nonhuman nature within it blinds us even further to the potential of stepping away from the assumption of metaphysical truths, blinds us to the dangers of continuing to rely on this sense of human dominance as a source of stability, and blinds us to the possibilities for human creativity and wellness in pursuit of a pragmatic project. Hans Joas ties the project of pragmatism directly to the project of sustainability as the work of revealing the consequences of human domination of nature:

> If Dewey's assumption is correct that classical metaphysics is a result of the attempt to dispel the risky character of everyday life by a philosophy of necessity and atemporal stability, then the rising power of mankind and the domination of nature are a historical precondition for the rollback of this metaphysics. The unintended consequences of this domination of nature which we contemporaries experience so intensely are able, however, rapidly to deprive one of the course of this insight. Then there arises a secondary covering up of the natural conditions and potentials of human creativity and the dangers of the human condition. Against such a secondary covering up, as it happened prototypically in Heidegger's philosophy, we have the project of pragmatism inspired by John Dewey and George Herbert Mead.
>
> (Joas 1993: 259)

Applying this thinking to the domain of planning, Robert Lake reflects this same understanding of the need to see new, unpractised possibilities for pragmatic truth in order to arrive at outcomes that hold more value for justice than our previous thought-crutches have offered:

> A planning process that passively (or "realistically") accepts the limits on action imposed by hegemonic structures, and that considers such structures as givens external to planning, both reproduces that hegemony and employs a truncated vision of justice restricted to just so much justice as is compatible with the requirements of a prevailing ideology.
>
> (Lake 2016)

A pragmatic, process-based approach to sustainability counters this tendency to rely on dominant assumed foundations and prevailing ideology by setting up a discursive playing field in which the societal discussion about what kind of world we want to live in can take place. In the context of discussion, the prevailing ideology that limits our work includes stories that sustainability is not sufficiently engaging an idea to compete with, say, health or wealth as motivator for behaviour change, that sustainability does not go far enough and always represents a poor compromise, that it is primarily focused on environmental considerations, not human lives, and that is has a tendency to accept scientific understandings uncritically and out of context. Such understandings must be cast aside if sustainability is to be understood as a concept that is supportive of justice. Different, more desirable, substantive narratives about sustainability exist, in rarefied academic discourse (Miller 2013; Robinson and Cole 2015). From a pragmatic perspective, these must be put to the test in an enriched public discourse. In this way, we can determine where the truth in sustainability can lie. Can it be a widely and personally engaging concept, can it can go beyond harm reduction and damage limitation to reach for regeneration and transformation, and how far can the concept broaden out beyond the environmental dimensions of sustainability and a narrowly realist view of science and technology? There is a gap in perception that dogs this conversation, time and again, because of the distinction that many make between what can be demonstrated to be true and what has moral (ergo subjective) value.

To James, no clear distinction can be made between a true idea and one with moral value, because determining what one believes to be true in a world without absolutes (as he believed the world to be) is a moral act. To quote James in trying to explain this notion to his absolutist critics:

> *The true … is only the expedient in the way of our thinking, just as the right is only the expedient in our behaving.* Expedient in almost any fashion, and expedient in the long run and on the whole, of course; for what meets expediently all the experience in sight won't necessarily meet all farther experiences equally satisfactorily. Experience, as we know, has ways of *boiling over,* and making us correct our present formulas.
>
> (James 1977b [1907]: 312, italics in original)

Thus, he understands it to be

> a matter of common observation that, of two competing views of the
> universe which in all other respects are equal, but of which the first denies
> some vital human need while the second satisfies it, the second will be
> favored by sane men [*sic*] for the simple reason that it makes the world
> seem more rational.
>
> (ibid.: 313)

That which is helpful, which serves a need in the long run or makes a difference to human lives, should also be that which we consider true because, so
long as we are able to recognize the connection, it would be incomprehensible
for us to hold a belief that does not serve a vision in which our needs are met
by the given world. The point, when extended to thinking about sustainability,
is that sustainability can only attain true meaning when defined by an interacting human community and that no human community would create a
concept of sustainability with such truth value that would be antagonistic to its
own generalized sense of wellbeing. James takes a radical stance here in relating
one's judgment of truth to one's determination of utility, promise, or value.
Moreover, in the pragmatic view, this never-ending process of verification of
truth-in-experience is humanity's only storehouse or referencing system for
morality, or justice. This process is our only available foundation and arbiter of
justice, and of sustainability.

In coming to understand James on this point, we need to grapple as well
with his notion of rationality. To James, a desire to find rationality in the world
makes his concept of truth sensical and sensible, because it makes human
judgment of truth and value predictable. Pragmatic rationality is an aspirational
rather than an analytical idea. In Hoch's (2007) view, we plan in order to
develop habits of rationality rather than imposing rationality on the practice of
planning. In other words, we plan not because we assume rational behaviour but
so that we might take more rational actions. Instead of divorcing means from ends
as in the rational comprehensive model of planning, we can think of pragmatically
rational planning as working on a "means-ends continuum" (ibid.: 21).

Compare this notion of rationality to that of "public rationality" proposed by
Robert Park, the founder of urban ecology within the Chicago School. Rationality,
to Park (1926), was a "form of communication, and one that is limited by one's
cultural community" (Bridges 2005: 3). An individual's rationality, thus, is culturally dependent. This understanding of rationality drastically shifts the positions of
economic or scientific rationality from their accustomed places, as universal
systems of communication, to systems that are just as situated in a particular
social organization and community as any other system of communication.
Pragmatic prescriptions for change emphasize processes that develop new relationships of trust, respect, and regular patterns of action, rather than specific
one-time products or outcomes. This provides a means to guide change in
desired directions while recognizing the indeterminacy and contingency of our

situations and knowledge. Its basis is the notion that human civilization ought to improve with the expansion of human inquiry and its technological products. As human experience "boils over," in James's evocative phrase, it lays new ground for improvement of rationality. And the process continues. In addition to supplying a means to act in the face of uncertainty, a pragmatic approach thus provides a certain level of optimism for sustainable development (Light and Katz 1996).

A number of contemporary environmental philosophers defend the potential of pragmatism to provide a sound philosophical basis for sustainable development. Prominent among these is Bryan Norton (1999), who argues that sustainability must be a pragmatic idea if it is to hold weight for the future of civilization. Pragmatism's orientation toward the future allows for the consideration of the needs of those who will come, alongside those of present generations, for the evolution of new life forms and ways of life in a way that facilitates the survival of a growing community of inquirers, and for human endeavor to improve our lot on the planet despite deteriorating odds. Further, sustainability makes sense through a pragmatic lens because of the value of the pragmatic concept of social learning: "it is open-ended and therefore requires social interpretation and experimentation before it can find expression in practical policies in response to practical problems" (Eckersley 2002: 53).

According to Norton (1999), any successful definition of sustainability must include a pragmatic notion of truth. That is, truth in sustainability must be thought of as transformation within a group, rather than as a movement to bring ideas into correspondence with an external reality. This conceptualization of truth involves social learning, understood as a continuous process of bridging the fact-value gap within a community whose membership grows over time. Norton's 'truth as transformation' means that truth should properly be judged in a forward-looking, morally relevant, transformative way. Truth is the reality that a community of inquirers will eventually agree upon voluntarily, at some undetermined future date.

In terms of an approach to urban sustainability, pragmatism therefore entails a commitment to searching within a community of people for insight into problems of action and inaction, for new habits to form in the face of uncertainty, and for a conception of truth based on social learning and as relational and forward-looking. These tenets continue to tie pragmatism to the work of James, whose philosophy was designed to reconcile the too-common divide between deep thinking and action. By developing habits of practically acting out our knowledge, James believed that people could be convinced of the utility of joining with the community of inquirers by contributing their unique perspectives to the ever-expanding fountain of common knowledge and experience of the democratic public. In terms of social justice, the goal for pragmatism is to determine the differences that our world formulas make to our recommendations for change, and the effectiveness of those prescriptions. Not least of the potential achievements of a pragmatic approach is the coalescence of a growing community of inquirers in a social learning process toward a truth that is constantly being forged.

The philosophy of pragmatism, then, provides a path to understanding and taking action on the project of sustainability. That path is process-based within a community of inquirers and thus in a way that demands that justice be served. The challenges to adopting a pragmatic approach, however, are monumental. Such an approach demands a commitment to the possibility of forming and expanding a community of inquirers, one that could, at least at some point in the future, represent an inclusive set of those invested in sustainability and justice outcomes for cities. This in turn demands a commitment to local democratic process, social learning, and the ability of such a process to overcome other, more dubiously motivated, compulsions to short-term and individual power. All of this, at a time when trust in process and in public participation in local governance has become associated with a dynamic of dread. Many potential members of a community of inquirers have become disenfranchised or radicalized by bad experiences or a lack of hope, feelings of persistent or intentional exclusion, and powerlessness. Others have built reputations, careers, and understandings of success around the intentional disruption and manipulation of efforts to engage, learn, and build alternative futures; they are intentionally constructing new barriers along the way.

On the other hand, continuing participation in the operations and social and sustainability work in our cities demonstrates that, despite all of these risks, many people still want to connect with others around issues that matter in the public sphere. Participation as a democratic exercise cannot be written off as a process entirely coopted by capital or corporate interest so long as citizens are demanding them and are voting with their heads and their feet for their continuation. The appeal of learning cannot be dismissed as incapable of standing up to the appeal of political power so long as learning processes continue to occur, in constantly new and innovating formats, drawing participants of their own accord, for a host of reasons, at a variety of costs and trade-offs. Nor can cities' pursuit of an ecological modernization agenda be passed off as subjecting a naïve vision of sustainability to the demands of capital in this light. Not so long as these efforts share the load of governance with citizens and other non-traditional actors, in new ways, injecting new possibilities for surprises, learning, and new engagements.

A critical pragmatic approach to the pursuit of urban sustainability and justice provides a means of reading the potential in the work underway toward urban transformation. It allows us to shed analytic light upon what motivates actors to get involved and what actors consider to be important as they give reasons to 'justify' their interventions in the political process, whether constructive or not. These are treated not merely as power-games, but as representations made and actions taken by urban citizen-inquirers seeking a productive engagement – one oriented toward a better truth. Within the cacophony of voices and actions being taken toward diverse visions of justice and sustainability in the city, a critical pragmatic approach gives us tools to recognize the possibility of a new model of justice and a new set of values and justifications that advance particular values for sustainability. The analytical approach of critical pragmatism is presented next.

## A critical pragmatic approach to recognizing urban sustainability in action

Developed initially by Luc Boltanski and Laurent Thévenot and their colleagues at the Ecole des Hautes Etudes en Sciences Sociales, critical pragmatism, or the pragmatic sociology of engagement, provides a tool for analyzing disputes involving multiple actors in liberal-democratic contexts. The approach has two dimensions. First, it supports analysis of the reasons that actors give for publicly supporting or challenging an argument or course of action in a given social context. Second, it supports interpretation of what seems good to actors engaged in a dispute. To date, research adopting the approach has dealt with disputes over environmental conservation in France and the USA (Lafaye and Thévenot 1993), the uses of scientific data in French policy-making (Thévenot 1997), the impact of cultural shifts and economic change on the French welfare state (Boltanski and Chiapello 2005 [1999]), the emergence of political alternatives to neoliberalism in the UK (Davies 2012) and more broadly (Du Gay and Morgan 2013), organizational analysis in Denmark and Britain (Jagd 2011), corporate social responsibility efforts by Australian global corporations (Nyberg and Wright 2013), urban redevelopment projects (Holden and Scerri 2015; Fuller 2012), global carbon markets and 'green' urbanism (Blok 2013) and local planning disputes in Australia (Scerri 2014).

Boltanski and Thévenot conceive of philosophers or public intellectuals as "grammarians," who purposefully facilitate the construction of "social bonds" by offering one model among many competing models "of legitimate order" (2006 [1991]: 19). They explain what prompted them to examine intellectual interventions:

> When one is attentive to the unfolding of disputes, one sees that they are limited neither to a direct expression of interests nor to an anarchic and endless confrontation between heterogeneous worldviews clashing in a dialogue of the deaf. On the contrary, the way disputes develop, when violence is avoided, brings to light powerful constraints in the search for well-founded arguments based on solid proofs [such as those grammar-like ready-mades that public intellectuals supply], a search that manifests efforts toward convergence at the very heart of disagreement.
>
> (ibid.: 13)

The work to reveal hidden efforts toward convergence begins with the identification and articulation of the different models of justice that people bring into play in particular disputes, in particular contexts. Models of justice allow people a means to order the relative worth of statements, objects, other people, and even themselves. Models of justice designate different individuals or figureheads as important or worthy, who then reinforce the legitimacy of the order: "By their presence [worthy people] make available the yardstick by which importance is measured" (ibid.: 141). Adherence to a model of justice bridges both the cultural and political roles that people may play in public life.

Importantly, there is a performative, interactive, and context-dependent aspect of a model of justice in addition to a linguistic component, observable in the process of dispute. This component is key to differentiating a pragmatic approach to analysis from a communicative approach which has become the dominant model used within participatory planning and democratic theory contexts. The thicker, pragmatic understanding of the communicative contexts that need to be understood and enriched for more democratic outcomes finds a root in Dewey's vision of radical democracy, as expressed by Bernstein (2010: 86):

> Ever since the 'linguistic turn' there has been a tendency for democratic theorists to focus almost exclusively on speech acts and linguistic procedures for adjudicating differences. But Dewey's vision of radical democracy is much thicker. It is not limited to deliberation or what has been called public reason; it encompasses and presupposes the full range of human experience. Democracy requires a robust democratic culture in which the attitudes, emotions, and habits that constitute a democratic *ethos* are embodied.

A critical pragmatic approach offers a way of partitioning out the identifiable perspectives rooted within complex and multiple interacting public debates. The analysis relies upon a small, but changeable, set of established models of justice, which are invoked by different actors in different contexts, and clash accordingly. The 'industrial model,' for example, can be approximated by old-time conservative values of industry and efficiency; an actor may choose to argue for the value of any action that favours sustainability, and may concomitantly denounce arguments that discount industrial model values. The 'inspired model,' another established model, in which value is associated with inspiration from above and beyond, such as religious or spiritual inspiration, would be another way of valuing and arguing for some kinds of sustainability shift. Each model will articulate a different definition of sustainability to suit its particular understanding of the ordering of value in the world and how this value can be advanced. These different models of justice could well face deadlock in understandings of value, and present incommensurable paths forward. On the other hand, if this mix of actors acts out of respect for the democratic value of the public sphere, they will seek to compromise in order to bridge the distance between distinct ways of understanding justice.

Nor is examination of models of justice in practice an exercise in linear replacement of one dominant model by another. When participants rise to demands to give reasons for their arguments and their actions, they call upon old and sketch out new understandings of worth, value, and fairness. They may win over adherents in so doing, who come to recognize these understandings as their own or to prefer these understandings to what they had previously held to be true.

In addition, compromises may be sought as middle ground between two or more operative models of justice. Far from being a route to mediocre results that meet no objectives well by aiming to meet too many diverse objectives, these compromises can represent key learning and social space that divergent actors create through their engagement. Iris Marion Young (1990) argued that compromise is essential for operating in an urban mode in a way that pursues justice. That is, we cannot possibly all understand one another, agree with one another in all (or even most) respects, but through compromise we gain the opportunity to live with our differences and move forward in spite of them. A compromise is an initial, democratic step toward progress in understanding and effecting ordinary justice.

Moreover, if reinforced over time and by more actors, the kinds of justifications that start out as compromises can solidify in public discourse into entirely new models of justice. If a compromise is subject to tests of coherence with existing valued structures and ideas, it begins to demonstrate pragmatic value to an increasingly inclusive set of citizens. It may then settle in public understanding as a new model of justice. As a model of justice, a notion would be more durable in public discourse than a contextually specific compromise.

In this book, we ask whether we can see evidence of a yet different kind of invocation of sustainability in public disputes, one in which a common grammar of disagreement is being constructed, across the different extant models of justice that can be recognized in other contexts of public disputes that may constitute an emergent new model of justice as sustainability. Do the expressions of sustainability constitute a particular identifiable formulation of the common good, a new understanding of virtuous people and roles, virtuous activities and objects, and their inverse? Is a critical mass of power and membership consolidating in different contexts, constructed by the notion of sustainability as a model for ordering the world of goods and bads? As a model of justice, sustainability needs to be valued in the public sphere because it contributes to the common good, and also because actors judge it to be valuable on their own terms. 'Sustainability' would have to be not only something that 'we all want,' but something that different actors in public life can and do see as a source of economic gain, efficiency, or prestige. If this is indeed happening, then we may be witnessing the awakening of a different moral ordering of Western culture, a new way of understanding value and status, an alternative grip on the world and what it means to work and live within it.[2]

The notion of creating a new sustainability model of justice seems all the greater a challenge because it would seem to require the intentional and nearly simultaneous act by many of displacing their pre-existing beliefs about justice with new ones. This is an uncommon occurrence in contemporary society, considered limited to mystical events of epiphany, religious awakening, and the like. Such an intentional act of mass transformation, however, is only one of several paths possible for the societal adoption of a new model of justice oriented toward sustainability. Another possible path to change is the incremental route. This path is much less intentional, reliant more on the automatic substitution of bits and pieces of regular behaviours with new ones in response to different kinds and quantities of feedback, criticism and approval. Over time, the results of particular decisions, whether positive or negative for different individuals' understanding of their reputation, self and family advancement, and other personal goals, can change behaviour and produce gradually greater consistency in behaviour. When changed behaviour becomes consistent, we can consider it tantamount to a change in attitude and beliefs, the assimilation of a new model of justice.

For this understanding of what motivates consistency in behaviour, evidence of this fixing of a new model of justice, we borrow from the thinking of Howard Becker on commitment. While acts driven by commitment can be expressed intentionally, they can also sometimes be expressed by default. For example, while I may sometimes act expressly due to a commitment to sustainability, or justice, or some other goal, sometimes my actions in favour of these goals will be driven at a more automatic level by decisions I make to settle other scores, to correct the pain of other losses. The more I come to internalize my commitment, the more my actions in its favour may be driven automatically. I may only later realize that my act is part of a consistent package of sustainability-oriented behaviours. My justification may follow my action, but be no less legitimate and authentic for this reversal. Becker (1960: 38), referring to the power of commitment generally in society, explains:

> Each ... trivial [act] ... is, so to speak, a small brick in a wall which eventually grows to such a height the person can no longer climb it. The ordinary routines of living – the daily recurring events of everyday life – stake increasingly more valuable things on continuing a consistent line of behavior, although the person hardly realizes this is happening. It is only when some event changes the situation so ... that the person understands he [sic] will lose if he changes his line of activity. The person who contributes a small amount of each paycheck to a non-transferable pension fund which eventually becomes sizable provides an apposite illustration of this process; he might willingly lose any single contribution but not the total accumulated over a period of years.

In the pragmatic approach, we observe a range of possibilities, means, and depths of engagement in urban sustainability work on offer in the public sphere. Some are trivial, some are monumental. Some will differentially line up with different people's interests and expertise, whether they realize this fully or only dimly, and will attract their participation, centrally or peripherally. Each of these possibilities offers pragmatic value, to the extent that it exhibits some of the classic democratic values such as learning with a community of inquirers, engaging with diverse others in the spirit of a common goal, and exercising one's own effort in a direction that serves a lasting public purpose. Each may also offer some value to the individual participant, whether in the form of a service brought to the neighbourhood from which they can then draw, a small grant to make some aspect of daily life easier or more enjoyable, a new social or professional connections, or an opportunity to think through and connect a local public issue to a matter of some personal or social concern and thus to communicate better across borders. All of these acts might add to the structures of commitment to sustainability, and to the justifications that people will craft in order to maintain the contours of the structure. It may take a while before we realize the weight and the circumference of the table we are building. A pragmatic analysis can help to reveal these features.

## Notes

1   Historical descriptions of the development of pragmatism can be found in Thayer (1968) and Ayer (1968). The seminal essay introducing the philosophy was Charles S. Peirce's (1998 [1878]) "How to make our ideas clear." This essay was made popular by James (1977a [1898]), beginning with his essay "Philosophical conceptions and practical results."
2   According to Thévenot (2014: 521):

> A [new model of justice] thus has a chance of being established when a group of actors, relying on a stable world of mechanisms and objects, sees its power consolidated, in such a way that its members feel that they are in a position to demand exclusive recognition, and pride themselves on a specific contribution to the common good, without having to assert or even excuse the strength acquired in the sphere they excel in by undertaking other, more acceptable virtuous activities. They can then seek to elaborate for themselves, and get others to recognize, a value, a status, which specifically defines the way they have a grip on the world, and give it a moral dimension. It is only then that the work of theoretical formulation is carried out (formerly pertaining to moral and political philosophy, and today, in large measure, to the social sciences) which makes it possible to extend the validity of the values thus identified, and make them the basis for a new form of common good. To put it in the language of *De la justification*, worlds precede cities, as a moment in a process of reflexivity whereby a certain form of existence acquires a meaning, and a certain world equips itself with a coherence and a style.

## References

Addams, Jane. *Democracy and Social Ethics*. Urbana, IL: University of Illinois Press, 2002.

Agyeman, Julian, Robert Bullard, and Robert Evans. "Exploring the Nexus: Bringing Together Sustainability, Environmental Justice and Equity." *Space and Polity* 6(1) (2002): 77–90.

Agyeman, Julian, Robert Bullard and Robert Evans. *Just Sustainabilities: Development in an Unequal World*. Cambridge, MA: MIT Press, 2003.

Ayer, A.J. *The Origins of Pragmatism*. San Francisco: Freeman, Cooper, 1968.

Becker, Harold. "Notes on the Concept of Commitment." *American Journal of Sociology* 66(1) (1960): 32–40.

Bernstein, Richard. *The Pragmatic Turn*. Malden, MA: Polity Press, 2010.

Bioregional. (2016) "One Planet Living." Available at: www.bioregional.com/onepla netliving/ (accessed 18 August 2016).

Birch, Eugenie and Susan Wachter. *Growing Greener Cities: Urban Sustainability in the Twenty-First Century*. Philadelphia, PA: University of Pennsylvania Press, 2011.

Blanco, Hilda. *How to Think About Social Problems: American Pragmatism and the Ideal of Planning*. Westport, CT: Greenwood Press, 1994.

Blok, Anders. "Pragmatic Sociology as Political Ecology: On the Many Worths of Nature(s)." *European Journal of Social Theory* 16 (2013): 492–510.

Blühdorn, Ingolfur. *Post-ecologist Politics*. London: Routledge, 2000.

Blühdorn, Ingolfur and Ian Welsh. "The Politics of Unsustainability: Eco-Politics in the Post-Ecologist Era." *Environmental Politics* 16(1) (2007): 185–205.

Boltanski, Luc and Eve Chiapello. *The New Spirit of Capitalism*. London: Verso, 2005 [1999].

Boltanski, Luc and Laurent Thévenot. *On Justification: Economies of Worth*. Princeton, NJ: Princeton University Press, 2006 [1991].

Bridges, Gary. *Reason in the City of Difference*. London: Routledge, 2005.

Bulkeley, Harriet. "Cities and the Governing of Climate Change." *Annual Review of Environment and Resources* 35 (2010): 229–253.

Campbell, Scott. "Green Cities, Growing Cities, Just Cities? Urban Planning and the Contradictions of Sustainable Development." *Journal of the American Planning Association* 62(3) (1996): 296–312.

Connelly, Sean. "Mapping Sustainable Development as a Contested Concept." *Local Environment* 12(3) (2007): 259–278.

Cowell, Robert and Susan Owens. "Governing Space: Planning Reform and the Politics of Sustainability." *Environment and Planning C: Government and Policy* 24 (2006): 403–421.

Curran, Winifred and Trina Hamilton. "Just Green Enough: Contesting Environmental Gentrification in Greenpoint, Brooklyn." *Local Environment* 17(9) (2012): 1027–1042.

Dale, Ann, William Dushenko, and Pamela Robinson. *Urban Sustainability: Reconnecting Space and Place*. Vancouver: University of British Columbia Press, 2012.

D'Alisa, Giacomo, Federico Demaria, and Giorgos Kallis. *Degrowth: A Vocabulary for a New Era*. New York: Routledge, 2014.

Davies, William. "The Emerging Neocommunitarianism." *The Political Quarterly* 83(4) (2012): 767–776.

Dobson, Andrew. "Social Justice and Environmental Sustainability: Ne'er the Twain Shall Meet?" In *Just Sustainabilities: Development in an Unequal World*, edited by Julian Agyeman, Robert D. Bullard and Robert Evans. Cambridge, MA: MIT Press, 2003, pp. 83–95.

Dryzek, John. *The Politics of the Earth: Environmental Discourses*. 3rd edn. New York: Oxford University Press, 2013.

Du Gay, Paul and Glenn Morgan, eds. *New Spirits of Capitalism? Crises, Justifications, and Dynamics*. Oxford: Oxford University Press, 2013.

Eckersley, Robin. *Environmentalism and Political Theory: Toward an Ecocentric Approach*. Albany, NY: State University of New York Press, 1992.

Eckersley, Robin. "Environmental Pragmatism, Ecocentrism, and Deliberative Democracy: Between Problem Solving and Fundamental Critique." In *Democracy and the Claims of Nature: Critical Perspectives for a New Century*, edited by Ben A. Minteer and Bob P. Taylor. New York: Rowman & Littlefield, 2002, pp. 49–70.

Ehrenfeld, John. "Sustainability Needs to Be Attained, Not Managed." *Sustainability: Science, Practice & Policy* 4(2) (2008): 1–3.

Ehrenfeld, John. *Sustainability by Design*. New Haven, CT: Yale University Press, 2009.

Evans, Robert and Simon Marvin. "Researching the Sustainable City: Three Modes of Interdisciplinarity." *Environment and Planning A* 38 (2006): 1009–1028.

Finn, Donovan and Lynn McCormick. "Urban Climate Change Plans: How Holistic?" *Local Environment* 16(4) (2011): 397–416.

Fitzgerald, Joan. *Emerald Cities: Urban Sustainability and Economic Development.* New York: Oxford University Press, 2010.

Friedmann, John. *Planning in the Public Domain.* Princeton, NJ: Princeton University Press, 1987.

Fuller, Crispian. "'Worlds of Justification' in the Politics and Practices of Urban Regeneration." *Environment and Planning D: Society and Space* 30 (2012): 913–929.

Garrison, James. "Pragmatism and Public Administration." *Administration & Society* 32 (2000): 458–477.

Gearin, Elizabeth. "Smart Growth or Smart Growth Machine? The Smart Growth Movement and its Implications." In *Up Against the Sprawl*, edited by Jennifer Wolch, Manuel Pastor, and Peter Dreier. Minneapolis, MN: University of Minneapolis Press, 2004, pp. 279–309.

Gunder, Michael. "Sustainability: Planning's Saving Grace or Road to Perdition?" *Journal of Planning Education and Research* 26(2) (2006): 208–221.

Hajer, Martin. *The Politics of Environmental Discourse: Ecological Modernization and the Policy Process.* New York: Oxford University Press, 1995.

Healey, Patsy. "Building Institutional Capacity Through Collaborative Approaches to Urban Planning." *Environment and Planning A* 30(9) (1998): 1531–1546.

Healey, Patsy. "Creativity and Urban Governance." *Policy Studies* 25(2) (2004): 87–102.

Hoch, Charles. "Evaluating Plans Pragmatically." *Planning Theory* 1(1) (2002): 53–75.

Hoch, Charles. "Pragmatic Communicative Action Theory." *Journal of Planning Education and Research* 26 (2007): 272–283.

Holden, Meg. "Urban Indicators and the Integrative Ideals of Cities." *Cities Journal* 23 (3) (2006): 170–183.

Holden, Meg. "Social Learning in Planning: Seattle's Sustainable Development Codebooks." *Progress in Planning* 69 (2008): 1–40.

Holden, Meg. "Urban Policy Engagement with Social Sustainability in Metro Vancouver." *Urban Studies* 49(3) (2012): 636–651.

Holden, Meg and Andy Scerri. "Justification, Compromise and Test: Developing a Pragmatic Sociology of Critique to Understand the Outcomes of Urban Redevelopment." *Planning Theory* 14(4) (2015): 360–383.

Hopwood, Bill, Mary Mellor, and Geoff O'Brien. "Sustainable Development: Mapping Different Approaches." *Sustainable Development* 13 (2005): 38–52.

Jaccard, Mark, Lee Failing, and Trevor Berry. "From Equipment to Infrastructure: Community Energy Management and Greenhouse Gas Emission Reduction." *Energy Policy* 25(13) (1997): 1065–1074.

Jacobs, Michael. "Sustainable Development as a Contested Concept." In *Fairness and Futurity: Essays on Environmental Sustainability and Social Justice*, edited by Andrew Dobson. Oxford: Oxford University Press, 2006, pp. 21–45.

Jagd, Soren. "Pragmatic Sociology and Competing Orders of Worth in Organizations." *European Journal of Social Theory* 14 (2011): 343–359.

James, William. "Philosophical Conceptions and Practical Results." In *The Writings of William James*, edited by J.J. McDermott. Chicago: University of Chicago Press, 1977a [1898], pp. 345–361.

James, William. "Pragmatism and Radical Empiricism." In *The Writings of William James*, edited by J.J. McDermott, Chicago: University of Chicago Press, 1977b [1907], pp. 311–317.

James, William. "What Pragmatism Means." In *The Writings of William James*, edited by J.J. McDermott, Chicago: University of Chicago Press, 1977c [1907], pp. 376–390.

Joas, Hans. *Pragmatism and Social Theory*. Chicago: University of Chicago Press, 1993.

Kania, John and Mark Kramer. "Collective Impact." *Stanford Social Innovation Review* Winter (2011). Available at: http://ssir.org/articles/entry/collective_impact (accessed 18 August 2016).

Lafaye, Claudette and Laurent Thévenot. "Une Justification Ecologique?: Conflits dans l'Aménagement de la Nature." *Revue Française de Sociologie* 34(4) (1993): 495–524.

Lake, Robert W. "Justice as Subject and Object of Planning." *International Journal of Urban and Regional Research* (2016). In press.

Larrick, Steve. "A Living Systems Model for Assessing and Promoting the Sustainability of Communities." Paper presented at the 1997 Annual Conference of the Community Development Society, Athens, GA, July, 27–30, 1997.

Latour, Bruno. "'It's Development, Stupid!' or: How to Modernize Modernization." (2008). Available at: www.bruno-latour.fr/articles/article/107-NORDHAUS& SHELLENBERGER.pdf (accessed 18 August 2016).

Light, Andrew and Eric Katz, eds. *Environmental Pragmatism*. New York: Routledge, 1996.

Lorr, Michael J. "Defining Urban Sustainability in the Context of North American Cities." *Nature and Culture* 7(1) (2012): 16–30.

Mang, Pamela and William Reed. "The Nature of Positive." *Building Research & Information* 43(1) (2012): 7–10.

Miller, Hugh. "Why Old Pragmatism Needs an Upgrade." *Administration & Society* 36 (2004): 243–249.

Miller, Thaddeus R. "Constructing Sustainability Science: Emerging Perspectives and Research Trajectories." *Sustainability Science* 8 (2013): 279–293.

Miller, Mel and Ian Siggins. "A Framework for Intergenerational Planning." *Foresight* 5 (6) (2003): 18–25.

Norton, Brian. "Pragmatism, Adaptive Management, and Sustainability." *Environmental Values* 8(4) (1999): 451–466.

Nyberg, Daniel and Christopher Wright. "Corporate Corruption of the Environment: Sustainability as a Process of Compromise." *British Journal of Sociology* 64 (2013): 405–424.

Park, Robert. "The Urban Community as a Spatial Pattern and a Moral Order." In *The Urban Community*, edited by Ernest Burgess. Chicago: University of Chicago Press, 1926, pp. 3–18.

Peirce, Charles S. "How to Make Our Ideas Clear." In *The Essential Peirce: Selected Philosophical Writings*, vol. 2 *(1893–1913)*, edited by Peirce Edition Project. Bloomington, IN: Indiana University Press, 1998 [1878].

Pezzoli, Keith. "Sustainable Development: A Transdisciplinary Overview of the Literature." *Journal of Environmental Planning and Management* 40(5) (1997): 549–574.

Reed, William E. "Shifting from 'Sustainability' to Regeneration." *Building Research & Information* 35(6) (2007): 674–680.

Rees, William. "Ecological Footprints and Appropriated Carrying Capacity: What Urban Economics Leaves Out." *Environment and Urbanization* 4(2) (1992): 121–130.

Robinson, John B. "Future Subjunctive: Backcasting as Social Learning." *Futures* 35(8) (2003): 839–856.

Robinson, John B. "Squaring the Circle? Some Thoughts on the Idea of Sustainable Development." *Ecological Economics* 48 (2004): 369–384.

Robinson, John B. and Ray Cole. "Theoretical Underpinnings of Regenerative Sustainability." *Building Research & Information* 43(2) (2015): 133–143.

Robinson, John B. and James Tansey. "Co-Production, Emergent Properties and Strong Interactive Social Research: The Georgia Basin Futures Project." *Science and Public Policy* 33(2) (2006): 151–160.

Rydin, Yvonne. *Governing for Sustainable Urban Development*. London: Earthscan, 2012.

Scerri. Andy. "Comfortably Inhabiting Reality: Justifying and Denouncing Arguments in a Development Dispute in the Post-Industrial Gentrified Inner-City." *Space and Polity* 18(1) (2014): 39–53.

Schumacher, E. Fritz. *Small Is Beautiful: Economics as If People Mattered*. London: Blond & Briggs, 1973.

Shields, Pamela. "The Community of Inquiry: Classical Pragmatism and Public Administration." *Administration & Society* 35 (2003): 510–538.

Swyngedouw, Eric. "Impossible Sustainability and the Postpolitical Condition." In *The Sustainable Development Paradox*, edited by Rob Krueger and David Gibbs. London: Guilford, 2007, pp. 13–40.

Thayer, Harold S. *Meaning and Action: A Critical History of Pragmatism*. Indianapolis: Bobbs-Merrill, 1968.

Thévenot, Laurent. "Un gouvernement par les normes; pratiques et politiques des formats d'information." In *Cognition et information en société*, edited by Bernard Conein and Laurent Thévenot. Paris: Editions de l'Ecole des Hautes Etudes en Sciences Sociales, 1997, pp. 205–241.

Thévenot, Laurent. "Voicing Concern and Difference: From Public Space to Common Places." *European Journal of Cultural and Political Sociology* 1(1) (2014): 7–34.

Verma, Niraj. "Pragmatic Rationality and Planning Theory." *Journal of Planning Education and Research* 16 (1996): 5–14.

Verma, Niraj and HaeRan Shin. "Communicative Action and the Network Society: A Pragmatic Marriage?" *Journal of Planning Education and Research* 24 (2004): 131–140.

West, Cornell. *The American Evasion of Philosophy: A Genealogy of Pragmatism*. Basingstoke: Palgrave Macmillan, 1989.

While, Aiden, Andrew E.G. Jonas, and David Gibbs. "The Environment and the Entrepreneurial City: Searching for the Urban 'Sustainability Fix' in Manchester and Leeds." *International Journal of Urban and Regional Research* 28(3) (2004): 549–569.

Wilson, David, ed. *The Politics of the Urban Sustainability Concept*. Champaign, IL: Common Ground, 2015.

Young, Iris M. *Justice and the Politics of Difference*. Princeton, NJ: Princeton University Press, 1990.

# 3  Celebrating the city, for all the wrong reasons?

Global society has gone urban (Brenner and Schmid 2014). This means that as the global population shifts toward cities, not only demographic, or economic, or environmental but also deeper cultural changes are afoot. The early years of this century have seen urbanism take on a new meaning, in culture, economics, environmentalism, politics, and virtually every other sphere of human life that matters. A variety of opinions exists on the nature and value of this massive scale shift in human understanding of cities, urbanism, and their role in our human future. The shift has profound implications for the ways in which we conceive of our roles and actions, decide on behaviours to pursue and avoid, and justify our actions in context; and these in turn have implications for justice in our actions toward one another as we act in real circumstances in our public lives.

This shift has and will make a difference to people's lives and life opportunities. All who care about the prospects for urban progress thus have an interest in understanding the implications of this shift for justice, sustainability, and other common goods. Do contemporary expressions of the value of the city and the urban turn for humanity constitute a particular identifiable formulation of the common good, a new understanding of virtuous people and roles, virtuous activities and objects, and their inverse? Is a critical mass of power, membership, and habits consolidating around the world toward a different dynamic of argumentation, as people struggle to put the good and the bad in a different order than what they were raised to do, in light of observed injustices and unsustainability? These questions address the prospect of finding evidence in the public sphere of a commonly accessible reordering of worth, what is valuable about life, and mode of justifying behaviours and ideas.

In this chapter, we part ways with critical urban scholarship by finding reason for a richer, deeper grounding of progressive human endeavour in our cities, from Vancouver to Vladivostok to Venice. We are searching for a sustainability tied to justice within the contemporary, global pro-urban shift. To do this, we draw upon the tradition of pragmatism from both sides of the Atlantic, as discussed in Chapter 2, to assess the landscape of new forms of public disputes and social contexts in today's cities. Applying a pragmatic sociology of critique to understanding this new urbanism can help us to consider sustainable urbanism as a potentially emergent new model of justification; and appreciate its implications for justice and the common good.

Lives and lifestyles available to urbanites have varied hugely throughout time, and continue to be enormously disparate around the world. Still, throughout history and certainly across the highly differentiated landscape of contemporary cities today, cities have held and continue to hold both individual and social promise. Today, more globally than ever before, the city is the soundest bet on which to stake one's claim for one's own future. Although people are not all that make cities, urbanism still represents the common property of humanity, in terms of the stock and flow of knowledge, goods, welfare, and the common opportunity that these represent.

If urbanity is to hold meaning as a social frame, that is, as a guide to conditions where people can live good lives, and to encourage people to live well, then we need a moral or justice-oriented way of thinking about the city and our urban lives. If people are publicly advancing arguments that treat sustainability as part of a conception of just outcomes, and if these arguments are meeting the test of coherence in the actions that result, then we have the potential to witness the awakening of a different moral ordering of urban cultures, a new way of understanding value and status, an alternative grip on the world and what it means to work and live within it. Bits and pieces of these discourses abound, in the expectations heaped at the feet of cities, their citizens, and their leaders. They surround constructions of the common good, the way public arguments are waged, and the struggle over articulating and justifying the truth in particular situations. Within all of this is the substance of a larger social question, still inchoate, about the purpose of our cities at this moment of dire crises of injustice and unsustainability. We see this in the transformation of how people justify their decisions on matters of value. In Part II of this book, we will explore a few of these justifications in depth, namely: the meaning of authenticity, empowerment and engagement, and risk-taking (what pragmatist Thévenot, 2014, calls "inquiétude"). We will consider the way in which our under-standings of these may be in a state of flux, and the prospects for justice and sustainability. We begin the chapter with a review of the state of flux, and then consider its implications for justice and sustainability.

## What was lost, now is found

Writing in the *Journal of Urban Affairs* in 1996, from Greenwich Village, Robert A. Beauregard sounded an alarm: the city was doomed, and nobody even cared anymore. Not since Edward Banfield had written *The Unheavenly City* in 1968, argued Beauregard, had a scholar written about the city in such a way as to raise the neck hairs of scholars and members of the public alike. There was no enthusiasm for, nor celebration of, the city as "primary object of contemplation," as something so big and important that no one with any hope of forming a coherent picture of the world could afford to ignore. This was a problem, to Beauregard, because if we stop thinking and talking about cities, we lose the thread, not only of our urban history, but of how to live well, and better, amidst diversity of all kinds. To Beauregard, what was at stake was no less than the act of "widespread reflection on the moral obligations that constitute American society" (Beauregard 1996: 218). Walking us through trends in suburbanization, fading government and media interest in urban affairs, and the demise of the public intellectual, Beauregard warned that if we lose urban thinking, we also lose moral thinking.

Today, we see a different picture. We see a resurgence of urban thinking, in which nary an academic discipline, political advice-giver, nor popular writer has left the city uncontemplated. In contrast to the predominant assumption before about 2005 of "the fundamental non-urbanity of North America" (Baird 2015: 130),[1] so much contemporary writing starts with and engages with an urban reality. We recognize, along with Brenner and Schmid (2015: 154), that this resurgence represents a seismic shift in urban studies scholarship: "the intellectual foundations of urban studies are today being profoundly destabilized." We acknowledge that this shift comes from a particularly North American perspective, as recognized by Gleeson (2012). Still, this is our starting point: before, cities were afterthoughts. Today, exposed to so many different scalpels, syringes, and spotlights, the city is treated as some kind of solution to any variety of suffering.

For urbanist scholars, this is an exhilarating moment in time. At the same time, *it doesn't all add up*. We agree with Gleeson that many questions are left unanswered in this resurgence of interest and enthusiasm and recognize the potential for policy and intellectual regression embedded within this trend. But, we also offer a response here to the critical urban scholars whose work rejects the promise of this particular urban moment out of hand. Critical urban scholarship sees the current urban celebration as little more than a ruse, a "black-boxing" of the epistemological dynamics at play in the urban resurgence, a masked abandonment of prospects for any alternative reality to today's unjust, unsustainable, unfettered and inhumane cities (Brenner and Schmid 2015). Is this kind of capitulation the real root of the celebration? From a pragmatic perspective, we cannot agree.

Thinking back to the 1990s, what has been lost? Beauregard thought it was the city; but, in hindsight, it was everything. Beauregard blamed the fact that the American middle class had relegated the city to history, via suburbanization of middle-class ideals, politics, and interests as well as its locational preferences. In retrospect, this temporary urban anomie was not anything specific against the city. It was, really, a quintessential 1990s moment of postmodern loss of meaning.

Today, urban discourse is back, with no small swell of passion. Rare would be the informed person today who sees urban affairs as unimportant to their lives, although this was a complaint of Beauregard about the urban anomie of the 1990s. And the new urban discourse cuts across traditional intellectual, policy, and public sphere divides. It particularly flourishes in the form of the online magazine, thanks to Richard Florida's CityLab, the Guardian Cities, Citiscope, Next City, Copenhagenize, *Metropolis* magazine, and others. The city is news, not just a place where news happens; the city is a fascination for people with all kinds of interests. And the city is a place for street-level politics, once again. Cities offer the necessary preconditions for political battles: big spaces where attention can be focused, and critical thresholds of human energy can be surpassed; concentrations of people amenable to having their passions roused; good cellular and wireless network coverage; and the quintessential social cross-section, the inability of the process to be owned and controlled by any particular private interest. People are marching for the city once again, middle and upper classes as well as the dispossessed, and for causes and concerns that range from democracy to capitalism to global climate change.

How did urbanism get its groove back? We can look to Edward Banfield, in fact, for a clue. Banfield thought that the city had met its match and was on a path toward necessary death via two imperative forces. The first was metropolitan expansion, or suburban sprawl. The second was the steady flow of the poor into the city, and of all higher classes outward, making for a persistently change-resistant urban social make-up. By Banfield's logic, to the extent that policy could improve the lot of the poor, the poor would move to the suburbs, such that the hopeless would always be left behind as intransigent urban problems. History has proven Banfield wrong; but he was not blind. The metropolitan expansion that he saw as death to the city has morphed into a multiform expansionist metropolitanism, instead. We are sprawling out and resettling in at the same time. In both directions, what we create looks more like the city than like the suburbs of Banfield's day. And, a few generations later, that same sense of pointlessness of whatever policy interventions the state might throw in the direction of cities, the clear sense of our own disposability and interchangeability, have come home to roost in the middle class, too. It is not only the poor whose best option is to think only of the immediate present, and not the future. Famously, now, 99 percent of us are supplemental, and likely to be ignored.

In nearly every corner of the world, urbanization is an inevitability. Amidst a great many reasons to lose hope, a chorus of urban celebrants grasps the city as a reason to hold on to hope for a better future within urbanization. We will look at the various reasons presented to celebrate the city, in the words of the books that are paramount to this celebration. Following this, we move into the critical urban studies literature which rejects the notion of hope in the present-day city entirely. Finally, we counter this critique with quintessentially moral urban prospects: to build and bolster authentic identities, achieve a sense of empowerment, and a new valuation of risk and human diversity. These urban potentialities are missing from both the new urban celebration and critical urban scholarship, but are essential to recapturing the prospect of moral thinking in contemporary North American society.

## The books that mark the shift in urban thinking

From his vantage point in 1996 New York City, Beauregard laments that the urban discourse stopped flowing after 1968. He is, of course, aware of Manuel Castells, Jane Jacobs, Michael Sorkin, Richard Sennett, and Sharon Zukin, among sparse others, who were investigating urbanism in the 1970s, 1980s and 1990s.

Castells in particular, with his 1977 *The Urban Question*, brought the material and social components of the urban into conversation in perhaps the prototypical full-bodied approach to thinking and acting out urbanity. Beauregard's point is that these works were read and discussed in the compartmentalized margins of the urban academy. They certainly did not contribute to a public discourse about cities, what they needed, or what potential they offered society. Even here, such moral perspectives on the meaning of the city were in the minority.

The more dominant thread of urban scholarship during this period came from an economic development perspective of post-Fordism (Amin, 1994). In the transition to post-Fordism beginning in the 1970s, cities moved toward flexible accumulation, divorced their success from the success of the centralized welfare state, and began instead to focus on economic niche development and competition with other cities for global status and investment. Cities valued entrepreneurial activities that differentiated their spaces and wares, and became entrepreneurial themselves. The result in urban space, as reflected by critical urban researchers, was the valorization of "divided city spaces" in which "space is functionally and economically shared, but subject to an increasing social and cultural segregation" (Groth and Corijn 2005: 505). In this scenario, planning spaces with narrowly defined uses and users become the trend to track. This was because of the perceived ties of such efforts to urban image creation, control, and competitive edge. Morality had little to do with it.

Now, in the world that is held in common by engaged intellectuals, public policy advisors, and informed citizens, the city is back. Those who pine to understand and talk about the city, regardless of their academic training or political leanings, now have Mike Davis,[2] Alan Ehrenhalt, Richard Florida, Benjamin Barber, Jeb Brugmann, Leo Hollis, Edward Glaeser, Allan Jacobs, Bruce Katz, Chris Kennedy, Charles Montgomery, and Alex Marshall, Jonathan Rose and Jeff Speck, to name some of the prominent authors to appear since the turn of the new century. For a more academic audience, we also have Saskia Sassen, John Friedmann, Susan Fainstein, Warren Magnusson, David Harvey, Avner de Shalit, and Andy Merrifield. It is a large cast, reflecting a diversity of backgrounds, approaches, and perspectives. In their convergence on the prospect of the city for the human future, we have much to learn from them; together, they offer a way to understand our particular urban moment.

This pro-urban shift has affected every domain of social sciences research. Beginning with the field of urban economics, in 2002, Richard Florida opened the floodgates to understanding urban economies in a way that did not depend directly upon big-smoke factories or traditional growth machines, but upon something intuitively more locally democratic, and accessible: ideas. City governments and urban dwellers alike jumped at the opportunity that Florida's creative class thesis offered to recentre themselves and their interests in the world of economics, and to chart a course for prosperity via talent, tolerance, and technology. Jeb Brugmann, in *Welcome to the Urban Revolution*, explains the urban renaissance in terms of a wrestling match between the idealistic, utopian, formal plans and expertise and the "more basic urban hunger" (2009: 161) of the human spirit. Florida, of course, saw clearly the power of urban meshing of the capitalist side of this urban hunger with its more creative and experiential side.

Bruce Katz, in his book with Jennifer Bradley called *The Metropolitan Revolution* (Katz and Bradley 2013), tells us that "innovation begets innovation," making cities and local communities the "solution bank" of the world. Not far off from Harvard economist Edward Glaeser, who, in *The Triumph of the City* (2011) as well as his regular column in *The New York Times*, lauds cities for their many qualities that favour innovation, individual achievement, and gratification. Cities, to Glaeser, are to be valued as pools of talent and human opportunity of every variety, for the excellence that arises from the division of labour, and for the low transaction costs offered by concentration and diversity.

Marshall (2013) turns Jane Jacobs' arguments about the organic nature of cities on their head from an economic perspective: it is the policy and institutional framework offered by government that creates the conditions in which the organic material of the city – its personalities, businesses, entrepreneurial moments of opportunity – may grow. This is a story of self-reliance from a local political as well as a local entrepreneurial perspective, a DIY sense of urbanists not 'waiting for the cavalry' or letting themselves 'drift' into oblivion. The incredible resilience of cities as social institutions enables cities to carry not just their own jurisdictions, boundaries, and populations, but also their partners and institutions, and increasingly, the provinces, states and nations in which they sit. Katz (2014) calls it "metropolitan activism," driving political change in stark contrast to the stasis at the national scale. This activism consists of "affirmative energy, collaborative problem-solving and pragmatic purpose." This is also the way Benjamin Barber (2013), in *If Mayors Ruled the World*, sees it.

In the domain of political science, Barber's new urban perspective on political science takes this thesis to its political and economic extreme. Not far off from Barber's thesis, Avner de Shalit and Daniel Bell (2011) explain how cities inspire political identities, and create a new lexicon and code to characterize the possibilities from this basis. Warren Magnusson (2013) offers us the opportunity to "see like a city" as a clarion call for urbanism in the twenty-first century. Rather than see like a state, as is the conventional mode of political science thinking (Scott 1998), Magnusson contends that it would be more productive to consider the state as an urban entity in an urban world order. Echoing Louis Wirth, Magnusson argues that urbanism is a way of life; thinking and acting that over-determine other modes like modernism, capitalism, and industrialism. Politics in the urban eye is more subtle and less confrontational than expectations at the state level. Cities are self-organizing, whereas individuals, competing authorities, and institutions are self-governing. Together this mash-up of processes produces order that is always temporary, unpredictable, and local. It also produces interdependencies, unpredictable relationships, and daily encounters that can turn political, although they were not necessarily so contrived. We hear echoes of Castells, here, who, with his 1983 *The City and the Grassroots*, argued that resistance to capitalism has always been urban. Whereas societies have lost their ability to identify with the nation state, as too distant, too sinister, too corrupt; and whereas the suburbs offer nothing to hold onto, there being no "there, there," we cannot help but see ourselves and our fortunes bound up together with our cities.

The discourses of the 1980s tended to assume that, in the context of globalization, cities were no longer relevant as economic units, given the rapid flows of capital and financial information across national boundaries. However, the work of Castells and later Saskia Sassen brought home the idea that just as workers who facilitate these flows need to be housed somewhere, so do the binary data tapes documenting the continuous global flows of capital. Making this link between intangible flows of the global information age with physical built and ecological infrastructure of cities reasserts the role of the built environment and material fixes for the city. It further reasserts the role played by workers and information technology systems in operating and populating the global economy. Cities are fixed, and cities facilitate flows.

The study of urban design and engineering has also been affected by this shift. Christopher Kennedy (2011), in *The Evolution of Great World Cities*, focuses on what is considered to be most fixed in cities and reminds us of the essential, foundational nature of cities' underlying infrastructure, which determines their rise and fall. Allan Jacobs (2011) depicts how form matters essentially to the overall social and political make-up of particular cities, too. Kennedy's is a fascinating and compelling historical reinterpretation of the rise and fall of cities, from an engineering perspective with a growing ecological awareness. Jacobs writes a similar tale from a long career of impassioned work on the design of urban public spaces. To Kennedy (2011), infrastructure is everything when it comes to cities reaching maturity and keeping energy cycling within instead of channelling elsewhere. Jacobs takes a reflexive perspective on what local people have done to come to own the public spaces that urban design has created, reflecting on the possibilities of local people and local caring about their home city to invest in good urban infrastructure and form that, as part and parcel of the process, also develops strong urban social communities.

In urban sociology, finally, Saskia Sassen (1991) opened our eyes to the importance of particular world cities over nation states in increasingly globalized economic and political flows. Then, from an anarchist perspective, building upon Castells's (1977) take that the urban is a space of collective consumption, increasingly glorified and exclusionary, Merrifield (2014) hones in on the need to generate a populism out of the urbane 99 percent in order to rectify the inhumanity of the 1 percent in marginalizing and disenfranchising the rest of us. He represents those whose passions have been roused by the concentrated humanity in cities, via the Occupy Movement, the Arab Spring, and other aspects of human class-based struggle worldwide. He calls for "an understanding of what gives cities their frightening force and awesome grandeur … ambiguity and contradiction and conflict" (Merrifield 2002: 17). Urbanists like Merrifield draw from the work of Lefebvre and Harvey to see the city as a fertile site for change because of the position of an urban daily reality between the large and powerful institutions of the state and the informal rules of society – a position from which new rights can be asserted, powerfully (Kipfer 2009).

These books, their theses, and the passion with which their authors treat the city, offer clear and comprehensive reasons to celebrate the urban age. Cities are celebrated for the diverse varieties of hope offered by these varieties of urban economics, governance, civil and landscape infrastructure, social visions, and mobilizations for change. Far from the indictment of Banfield in the 1960s or the fears of Beauregard in the 1990s, the city today is an essential tool for prosperity, democracy, sociability, and human experience of the good life. Through the work of these writers and others, cities have demonstrated the many intersecting ways that we both need and desire them, deeply. This represents a monumental change in thinking whose impact will come to be felt across the social and policy sciences, not to mention the humanities, in years to come. But this does not explain *why* this has happened, why *now*, and it does not answer what difference it makes to those who *actually* think and act morally. By looking for evidence of how revaluing the city can contribute to the moral content of the city, we invoke the test posed by pragmatist philosopher William James at the turn of the twentieth century, who sought the "cash value" of an idea.[3] What is the practical utility of the city for urbanites? How does the city generate a different experience of the world, one that might promote an emergent urban ethics and morality aligned with both sustainability and justice?

## Trappings revealed by critical urban scholarship

Through the contemporary urban celebration, critical urban scholarship has raised serious doubts as to any notable value of the contemporary city for goals of justice and sustainability. Critical minds are adept at observing what looming or lingering problems exist in the contemporary urban celebration. In what follows, we examine the way that today's chorus of prominent urbanists presents the city's opportunities and hides or discounts its problems. In what follows, we zero in on three critiques, found in critical urban studies literature, that present traps into which the urban celebration can often fall. They are, in turn: (1) the local trap; (2) the empowerment trap; and (3) the community trap. We who care about prospects for work toward sustainability and justice in today's cities should value these critiques for the warnings and starkly different viewpoints they provide. Taking these critiques into account allows us to give credence to a fuller range of our urban experiences and the meaning we make of them. If, after considering these critiques, we retain a sense of affinity for the promise of the city, we can offer greater confidence in our pragmatic commitment to the urban potential for sustainability and justice.

In the remainder of this chapter, I will present the case for each of these three traps. I will build upon critical urban scholarship to conclude that, in terms of the local trap, what we celebrate about cities may often be confounded with what is simply local, and may not transcend a preference for a particular inscribed set of geographic boundaries. And, from the perspective of the empowerment trap, many attempts to increase social and political engagement in cities may not be motivated by authentic democratic commitment but rather by a fearful sense that most people need to be subjugated to ensure core urban functions are filled. And, referring to the community trap, what is celebrated as grassroots social mobilization and community organizing within cities may in fact more closely represent the incipient opening up of new terrain into which market forces can move and establish themselves to supplant voluntary and extra-market activity. Together, these traps represent clear reason for doubt and skepticism about the real possibilities for justice and sustainability in cities today. We consider these traps not to dismiss the urban potential, but to motivate, redirect and reinforce the need for conscious urban work and attention, much more than simplistic celebration.

### The local trap

Within the urban renaissance is a remaking of cities as local, parochial places in which people can dig in to local life. The localized urban ideal presents a picture of genuine lives being authentically lived. This is possible by maintaining a modest circumference of life that respects durable physical limits to how far people are willing and able to spend travelling on a daily basis. This remaking of complete, compact cities and urban neighbourhoods is made possible by the shift from industrial to post-industrial economic drivers in the developed nations. This has reduced the health and wellbeing objections to placing homes, shops, and schools near job centres as these job centres are now typically less polluting and less out-of-place. New fundamentals of urban design have been introduced or reintroduced, harking back to a simpler, friendlier time, when local concerns and local offerings dominated people's daily lives. These fundamentals have spread to all corners of the aspiring urban celebration to create mixed-use, more completely 'liveable,' neighbourhoods.

The liveable urban design aesthetic, often referred to in the United States as neo-traditional design, or the New Urbanism, reimagines a post-World War II American urban streetscape, with narrow streets to keep cars moving slowly, sufficiently wide sidewalks to teach children to ride bikes and skip rope, stoops and front porches for adults to chat and newspapers to be tossed, trees and ravines to attract little birds and large frogs, small and closely spaced single family homes or townhouses, no front driveways or yards, and all of it within walking distance of a high street with daily shops and services, perhaps with apartments on top of the shops there. These neighbourhoods are valued for the options they bring in terms of mobility, housing type, employment, education, recreation, and ecological and energy efficiency, compared to traditional suburban neighbourhoods. "Complete communities," "walksheds," and "5–20-minute communities," are the intended result: neighbourhoods that have jobs, shops, and services within a 20-minute journey of most homes, via walking, cycling, and/or public transport (Lowe and Giles-Corti 2015). While a minority of urban districts actually achieve this, the aspiration to meet these design and land use goals has nonetheless become standard and accepted in most urban planning contexts.

The value attributed to such design practices goes well beyond individual benefits for those who adopt the lifestyle offered in these places. They incorporate place-based social value that is asserted, also, for neighbourhood cohesion, social capital, community-building, and even overall health and happiness. They are building upon the insights of such visionary urbanists and urban observers as Jane Jacobs, William H. Whyte, Allan Jacobs, and more recently, Jan Gehl and Bjarke Ingels. Ingels, for example, refers to his style of urban architecture as based upon an evolutionary process of society. Using an experimental approach, in which he and his group expect some prototypes to fail so that others may succeed, Ingels crafts a "pragmatic utopia" in which the perceived conflicts in demands for the built environment can be solved through more creative design thinking. In this way, for example, the perceived trade-off between ecologically suitable and economically profitable built environments can be overcome "by tying conflicting interests into a Gordian knot of new ideas" (BIG 2016).

The social and physical connections in this body of research and practice are now encapsulated in the notion of placemaking. Placemaking, a term coined by New York-based Project for Public Spaces (PPS) founder, Fred Kent, and now in use throughout the urban planning and design professions, ties urban design, and particularly the close interaction between private and public space, explicitly to community-building as a social goal. A "placemaking" approach "helps citizens transform their public spaces into vital places that highlight local assets, spur rejuvenation and serve common needs" (Project for Public Spaces 2015). Through engaging in this process of placemaking, Project for Public Spaces contends, strong urban communities are built and maintained.

On the other hand, critics have argued that no necessary connection exists between the urban design strategies of placemaking and the imputed outcomes of social health and harmony. They point out that living in so-called liveable neighbourhoods entails trading off more private space, often without cost savings. This relationship appears most starkly true when comparing the cost of housing directly in a dense urban neighbourhood with the cost of housing in a suburban one. Nor can new urban residents expect to spend less on consumer items; shopping at the neighbourhood high street will almost certainly mean spending more for food and household items than shopping at the big box store on the outskirts of town. In some of the most successful examples of such neighbourhoods, the cost of housing and the cost of living become so expensive that the motivation to rent out one's home to others becomes incontrovertible; this choice has implications in turn for the social capital and make-up of the neighbourhood. Those living in these neighbourhoods typically still own a vehicle, though they may drive it less; so they are paying for this and they have less space, or less convenient space, in which to accommodate it (Freytag et al. 2014; Westerhoff 2015). The argument is stretched further into the realm of opportunity costs of living in a new urbanist neighbourhood: if the bulk of residents have similar educational and employment aspirations, within creative class industries, for example, this will make for a more competitive and ultimately tougher outlook for those seeking scarce spots in universities and desirable companies. For a suburbanite, taking into account the trade-offs of more expensive housing, smaller private spaces, limited space for one's vehicle, expensive shopping, and more competition for school and jobs, the choice of living in a new urban neighbourhood is not so clear-cut (Marshall 2001; Rybcyznski 2010). Even the promise of energy efficiencies in this kind of design has significant caveats based on social practices within and spending habits outside the home (Sussmann 2012).

The orthodoxy that has emerged represents a rejection of the cruelty and indifference of the comprehensive rational model in planning and modernism in design and architecture (Sandercock 1998). Take, as emblematic of this brutal older model, that still lingers, the Charter of Athens (Congrès internationaux d'architecture moderne 1946 [1933]), produced by the IV International Congress of Modern Architecture, chaired by Swiss architect, Le Corbusier. Addressing what was perceived as the "chaos" of uncontrolled urbanization in the cities of the day, the Athens Charter established a reasoned set of principles for urban planning meant to improve urban quality of life. Among 94 points, many of which would be recognized as having relevance today, we also find the core points against which the new urban design rebels, namely:

1   [The] segregation of dwellings is sanctioned by custom, and by a system of local authority regulations considered quite justifiable: zoning.
2   Buildings constructed alongside major routes and around crossroads are unsuitable for dwellings because of noise, dust and noxious gases.
3   High rise apartments placed at wide distances apart liberate ground for large open spaces.
4   Unsanitary slums should be demolished and replaced by open space. This would ameliorate the surrounding areas.
5   Street widths are insufficient. Their widening is difficult and often ineffectual.

Segregated housing, slum removal practices, the creation of large housing blocks set back from the street on green lawns, wide and fast separated freeways and parkways, are all emblematic of the roots of unsustainable cities of the twentieth century. The new urban design orthodoxy prioritizes the opposite principles: favouring inclusionary over exclusionary zoning, mixed housing types, gradual revitalization of decaying neighbourhoods rather than their bulldozing and erasure, streets that are narrow and dotted with buildings, stoops and pocket parks, and gardens, in order to encourage non-automobile use.

Accompanying this urban design agenda is an urban economic development agenda aligned with Richard Florida's (2002) notion of the creative class. The creative class agenda is almost indistinguishable from the green capital economic agenda (Fitzgerald 2010) and stakes its hopes for economic development in non-polluting, non-resource-intensive fields that demand not industrial zones segregated from housing but smaller spaces mixed in tightly with small homes inhabited by educated people, expressing themselves through their work, enjoying urban amenities and street life. In new urban design thinking, this is a clear win–win: better urban design for better quality of life, and better economic development opportunities to boot.

This new urban design has been further justified in terms of improving human health through environmental design. As non-communicable diseases such as cardiovascular disease, cancer, mental illnesses and Type 2 diabetes overtake infectious diseases as the leading cause of disability and death globally (World Health Organization 2011; Kent et al. 2012), urbanists' fingers point at design that results in land use segregation, sprawl, high rates of automobile use, poor air quality, and poor healthy food options (Capon 2007; Rao et al. 2007; Barton et al. 2009; Dannenberg et al. 2011). People living in these environments exhibit inadequate physical activity, inadequate social activity particularly within their neighbourhood, spend inadequate time outdoors, eat unhealthy food, experience social isolation, overweight and obesity, depression and other mental health challenges (Frumkin 2002; Gebel et al. 2009; Ewing and Cervero 2010; Cannuscio and Glanz 2011; Sallis et al. 2011; Giles-Corti et al. 2012; The Healthy Built Environments Program 2012; Saelens et al. 2012).

The new urban design strategies purport to achieve healthier, happier, more balanced ends through an emphasis on more local living. Attractive, safe, and nearby green and open spaces encourage people to spend more time outdoors, and to be more active; when these are public open spaces, this further encourages time spent socializing with neighbours (Sallis et al. 2011; The Healthy Built Environments Program 2012; Koohsari et al. 2015). Reduced commute times and shorter travel times for daily household errands also mean more time is available for these activities, and for the time it takes to build meaningful connections with one's family and household, with others in the neighbourhood, and with one's local, nonhuman environment, particularly through gardening and patronizing the local farmers' market (Frumkin 2002; Hartig et al. 2003; Gebel et al. 2009; Francis et al. 2012).

The move to create alternative ways of thinking about and living in local places is a valuable way to break out of the continuing legacy of modernist thinking. But as valid as these designs and claims of their impacts may be, they too have problems. The locally active neighbourhoods created by the new urban design orthodoxy often lack the ethnic diversity they would need to reflect the diversity of the surrounding city. More often than not, revitalization efforts force previous residents out as surely as the bulldozers have done before them. The new housing created, though it may be mixed in form and in proximity to streets, travel alternatives, and shops, may still only serve a narrow sliver of the economic diversity of the city. The creative class economic opportunities may further narrow this slice. As such, they represent affluent enclaves rather than emblems of the quest for urban justice and sustainability. In supporting locally grown produce via urban farming strategies, these residents may fail to support, understand, or even consider the need for farmers in traditional farms in the exurbs to have a dependable market for their wares, for example. In a more sinister tone, a number of urban critics would hold that this approach, when considered in terms of its outcomes, can best be described as implementing "serial forced displacement" (Fullilove 2015: 83) and a "blatant politics of exclusion" (Cruz and Forman 2015: 40). From a political perspective, it has been accused of being instrumental to a political state of "disaffiliation" with significant political interest and engagement (Castel 1995). This is the politics of the local trap.

Mark Purcell develops an argument against what he calls the "local trap" – a cognitive error made by progressive thinkers "in which the local scale is assumed to be inherently more democratic than other scales" (Purcell 2006: 1921). A person living a life of walkability, who shops and plays and relaxes in the public and commercial spaces of the home neighbourhood certainly has more opportunity to develop relationships with people there than someone who slips into the car as soon as they leave their home, still within the private confines of a garage, and makes little use of any public or commercial space in the neighbourhood. The range of ways in which the liveable city dweller will engage with their neighbourhood, and the range of values that they will come to attribute to their home, neighbourhood, and community, will be broader. A liveable neighbourhood dweller might express different values for their home than what is typical for a suburban dweller: a safe investment, place of pride, and sanctuary in one's private residence. The liveable neighbourhood dweller might share these, and add on a series of other values associated with the community and social spaces surrounding their home, because they would interact within these spaces on a regular basis, or at least intend to. The complete community may offer more local economic and social linkages, and perhaps more opportunities for understanding local ecology, even local governance processes, just by force of the regular habits of daily life. The path from one's front door to the transportation mode one uses to get to work or school, the path to collect one's mail, one's children, to walk the dog, tend the garden or seek physical exercise, the path to buy milk, butter, and bread – a liveable neighbourhood forces the use and appreciation of public and community spaces and local social interaction within these daily routines, while a traditional suburban neighbourhood does not.

What Purcell signals is that this public visibility of life routines does not ensure any particular political or ethical orientation toward the public in its political, rather than simply design, sense. There is no direct and proximate path to receptiveness to social difference within any given local neighbourhood design and land use mix, although a neighbourhood's design choices may well offer new opportunities and social rewards for spending time in public and community spaces. Design for local liveability does not have any necessary or explicit relationship with tolerance, respect for others or for one's environment, with a politically or socially open attitude, or with any particular normative stance related to either justice or sustainability. Although E.F. Schumacher (1973) gave rise to the movement toward local and liveable economies with his mantra, "small is beautiful," the beauty of many liveable neighbourhoods may be no more than skin deep. Liveable neighbourhoods can house those interested in, respectful of and committed to preserving nonhuman nature and those whose attitude toward nonhuman nature is fearful, indifferent, or antagonistic. They can be home to those who value the sanctity of private property above the rights of newcomers to establish a safe and happy home in a quality environment. Those who spend time and disposable income on locally produced goods and foods, riding in bicycle lanes, bird-watching in nature ravines and planting in urban forests can hold polar opposite positions on the question of who (and what other species and habitats) hold a right to the city, let alone to their neighbourhood. Moreover, many will purport not to hold any position at all on these deep questions of sustainability and justice, but will nonetheless exert decision-making power at critical political moments, sometimes arbitrarily, bringing about serious exclusionary consequences, or the inverse. Such decision moments arise when a new development is proposed, or a school upgrade denied, when a neighbourhood grant program passes over the neighbourhood for another, or some new neighbours take on new tenants, or a different religion, or new aesthetic preferences. When this happens, the arguments and decisions that follow can surprise even those individuals articulating them.

A preference for finding liveability at the local scale is, in this view, an example of misplaced scalar determinism. The local-scale orientation is taken to determine a just and democratic political orientation, as well as a sustainable set of relationships between people and nonhuman nature. Those celebrating the contemporary city on this basis are seen to have fallen into the trap of making assumptions about the city as a political, social, and environmental entity based upon an understanding of urban design logic at the local, complete community, spatial scale. Not only is this leap empirically unfounded, it can actually backfire. Assuming that local alternatives are always more just and sustainable because they are more "beautiful" can close off potential opportunities for building alliances, understanding different contexts and solutions, and sharing agendas across scales. Scale-jumping nonetheless has its own potential, such as when a local-specific effort links with larger states, provinces, and nations in which the local is embedded, with other places around the world, and with rural areas. Falling into the local trap can drain political energy from a place, by tempting solution-seekers away from specifying the legitimate normative goals which lie beneath the proxy of the local. Local-scale arrangements – and Purcell singles out arguments centred on the right to the city in particular – are neither necessary nor sufficient for good outcomes. Instead, taking the good of the local scale for granted may actually work to further the social divide. That is, privileging opportunities to build a sense of urban-specific authenticity may wilfully exclude broader and non-geographically defined goals and concerns.

In summary, for all the renewed attention to placemaking, the quality of a city cannot be reduced to questions of physical design alone – it is also a figment of human diversity and history, the urban civic culture, the community that expresses itself politically, or relates itself to broader geographies and communities, all of which needs cultivating. Urban public spaces open to and actually accommodating an array of users and uses in a wide range of configurations, are as key to their quality as the material infrastructure of the places. Whereas to an advocacy organization like PPS, the spatial dimensions and design features of neighbourhoods are key to their public nature, equal and adequate creative attention is needed to address the ideological and political ideas that may not be overtly visible in public space, but may manifest control and oppression just the same.

### The empowerment trap

The empowerment trap refers to a blind spot among those celebrating the contemporary city, in which any and all forms of empowering citizen participation and public voice in local decision-making are considered valuable to advancing democratic values and outcomes. The principle of subsidiarity is often invoked; that is, the soundest decisions are made at the scale most local to those most directly affected by the decision. Along with this is the notion that the cynicism, lack of trust, and anti-social behaviour exhibited by citizens with respect to national politics vanish at the local scale, where the clear and positive results of local engagement are self-evident (Barber 2013). *Cities act*, as proclaimed by mayors in opting to join the C40 cities initiative for carbon-neutral cities and so to make "a commitment to work collaboratively to address one of the greatest economic, social and environmental challenges of our time" (C40 Cities 2011).

Those particular mayors are referring to the particular challenge of climate change. Similar pronouncements have been made in favour of urban leadership on a host of other major challenges, from justice to resilience to sustainability. The general message is that leadership is wasted on the nation state. Cities have the right scale, the right proximity, and the right mix of motivations to make change happen. Breaking this perception down further, while today's city leaders receive considerable respect and reverence for their strong leadership qualities, urban decisiveness and progressive action are also attributed to other people ready and willing to mobilize for their cities. City leaders and their citizens, working individually and through groups, coalitions, corporations, affinities, and identities, form a natural and formidable force for empowerment, action, and progress.

At least, so this version of the story goes. Deliberative planners and sustainability advocates claim a natural and necessary connection between planning for sustainability and a broader public debate that redefines urban planning "expertise" to value the everyday knowledge of urbanites (Healey 2007; Holden 2011). This is often argued to be either natural or essential for three reasons. First, because of the integrated nature of sustainability planning, which demands a range of expertise. Second, because of the need for many diverse perspectives to weigh against the monopolization of efforts or effects by any given interest group. And third, because of the unprecedented nature of seeking sustainability in cities, demanding a more exploratory, experimental, as opposed to rational and linear approach. Promoting sustainability requires transformative governance work which Healey characterizes as *a messy back-and-forth process of multiple layers of contestation and struggle*. On the one hand, the aim is to create bold visions shared by multiple stakeholders. On the other hand, the work involves making decisions in order to move forward (Healey 2009).

Many urbanists refer to this transformative governance challenge in terms unrelated to the challenge of sustainability. Emblematically, Barber (2013: 5) claims that the "challenge of democracy in the modern world has been how to join participation, which is local, with power, which is central." Politicians are negotiating deals and arrangements that have the result of reducing public freedoms and public space for local residents, while residents are left with little sense of right to stake a claim or opportunity to voice a claim. Repeated ad infinitum, these dynamics widen spatial and opportunity disparities between the city of wealth and the city of poverty, the city being exploited and the city being rewarded, "millions of empty buildings and millions of people without a decent place to live" (Zarate 2015: 26).

Some cities have responded with a view to altering this dynamic, by instituting stronger, broader, more intensive participation and consultation protocols and checks on accountability of decisions to this participation. Attempts include that of the City of Portland (2010) to institute measures of the impact and uptake of citizen engagement initiatives on local government policy and action, which must be compiled by staff, alongside accompanying innovations in the engagement processes themselves. The aim is to create the conditions for deliberation as an institutionalized component of accountable partnerships. Under these conditions, the possible justifications for decisions and actions are discussed in public before the decisions and actions themselves are determined. From a democratic theory perspective, it follows that better policies are crafted (Dryzek and Niemeyer 2010). The value of understanding the role of public argument in the production of policy learning toward urban sustainability has also been noted by Bulkeley (2006).

Despite this promise of participation as creating the transition to the just and sustainable city, the ideal of local empowerment has traps. One major trap lies in the globally dominant trend of urban exclusion and divided cities, which participatory exercises do little if anything to alter:

> The conditions and rules currently present in our societies are globally condemning more than half of the world population to live in poverty. The inequalities are increasing both in so-called *developed* and *developing* countries. What real opportunities are we giving to young people if, according to the U.N., 85 percent of the new jobs at the global level are created in the "informal" economy?
>
> (Zarate 2015: 27)

For many of the world's urban slum dwellers, although they constitute absolute majorities of the urban population in some countries and account for the largest share of population growth in numerous others, opportunities to participate in formal urban decision-making will never present themselves (United Nations Human Settlements Program 2003). Their only opportunity for political empowerment may come through local uprisings, demanding concessions, rights, and resources with a backing of threat, rather than a backing of due process. The threat of disruption and violence may win this growing swath of urbanites more meaningful improvements to their daily lives than submitting to due process ever could (Das 2015).

Often, when a discursive process does occur between citizens and decision-makers, its outcomes demonstrate only the barriers between people and urban development decisions, with a predictable outcome that offers nothing by way of tools to remove these barriers (ibid.: 91). This echoes an earlier argument by Young (1996: 123), that deliberation in practice privileges arguments oriented to winning, a form of expression suited to well-educated and well-off men. Indeed, in processes oriented around decision-making, participants may only participate seriously if they believe it will influence decisions (Fung 2003) and participation tends to privilege those aspects of the process that are endowed with decision-making power (Kanra 2007). On the flip side, power-seeking behaviour within a participation process brings no real commitment to changing the nature and goals of the partnership among decision-makers. If participants are not in the process with a view to changing political dynamics, and maybe even changing their own minds, the process is likely to remain a decorative flourish that has little or no bearing on policy outcomes (Hertin and Berkhout 2003; Swartling et al. 2007).

Even more emphatically, the nature of public participation's effects on change in urban policy and urban form does not empirically demonstrate a necessarily positive relationship when it comes to sustainability (Holden 2011). The ambiguous results of attempts to engage democratically in debate about urban development spring from the idea that "liberty is the 'right to be left alone,' a private dream devoid of social responsibility" (Cruz and Forman 2015: 40). So popular is this sense that it can even be found among public participation professionals and participants. The logic goes like this: somehow, at the end of the day, we will leave these frustrations behind, because our participation in local democracy now will have earned us the right to be left alone. A dynamic of dread of public participation has evolved, in which members of the community may be radicalized by past experiences of disenfranchisement, feelings of persistent or intentional exclusion, and feelings of powerlessness. A small group of people may be driven to repeated engagement only in a staunch oppositional role, only to exercise their predetermined will to the extent of their ability. Momentum may be generated in a process at one moment, but then abandoned or drawn down the next. The will to engage and the will to empower may falter on either side. This can leave municipal staff and decision-makers on the defensive, feeling beleaguered by an aggressive and unreasonable small segment of the public.

Many conclude that the impact of public participation is perverse. Others hold that while public participation may permit perspectives held by participants into the channels of power, where they may have greater impact, it has lost whatever potential it ever had to stimulate learning among participants and process designers (Kanra 2007). Whatever impact participation processes have on decision-making, the process also shifts participants' mode of engagement toward strategic calculations of winning power and influence.

In a striking book that turns the relationship between empowerment and disempowerment on its head, Cruikshank (1999) characterizes this process of urban empowerment as insidiously eroding local democracy, and further consolidating power and capital. She explains the duplicitous role of power in government-led efforts that are put forward as related to citizen participation and empowerment. Whereas dominant pluralist political science assumes that the absence of visible domination is a good sign for freedom, Cruikshank looks deeper for signs of power being exercised without being obviously evident. Drawing from Foucauldian theory, she argues that efforts which ostensibly intend to empower others instead have the inverse effect, covertly desired by those who design them, of winning more people over to the necessity of the subjugation of many for the overall functioning of society. There are all kinds of ways in which power is exercised against people who are seen to be participating and engaged in a variety of citizen empowerment exercises. What engagement teaches formerly disempowered citizens, primarily, is that inequality and the silence of the majority are crucial elements of our system of governance: "What was invisible was not an act of power or the interest of the powerful but the silent consent of unequal parties to their inequality and to the system of government more broadly" (Cruikshank 1999: 15).

### The community trap

The third trap of the urban celebration is the community trap. Where the local trap blinds urban celebrants to the opportunities they may be missing in focusing design and living intentions at the local scale, and the empowerment trap masks the disempowering implications of local government decision-making, the community trap operates upon cities' assertions of social and community-building potential. There exists, within the urban celebration discourse, the notion that today's urban rediscovery and reorganization create a new version of the social collective within the city. This new urban collective manifests in aspects of the sharing economy, social media networks, a vibrant public and street life, a healthy and active civic sphere, and organizational life. The excitement to see newfound interest in community life is understandable. At the turn of this century, the "commons" was considered to be a vestige of an earlier time and system of social organization that was "destined to disappear in the face of modernization" (Agrawal 2002: 42). Graham and Marvin (2001) set the tone for the thinking of the millennium, namely, that globalized capitalism in the form of neoliberalism was grossly intensifying the fragmentation of urban landscapes into privatized enclosures of structures, infrastructures, and services. The ultimate conclusion of this process would be the splintering of all common spaces into fragments of their former selves, each dedicated to specific individualized purposes, with specific calculated values.

The tide has now turned. The life to be lived is back in the public spaces of community life, sharing in reasoned and bounded ways with one's circle of friends and neighbours. Correlations for this new and enhanced urban community life are offered with respect to reduced crime rates and the sense of safety that comes from better urban design that features "eyes on the street" and eliminates dark corners and blind alleys. As crime rates go down, mental health improves (Stafford et al. 2007), individual physical functioning improves (Ross and Mirowsky 2011), and people are more willing and prepared to engage with their neighbours in public. This provides a compounding effect, as this more neighbourly engagement spins off further positive psychosocial effects, including an increased sense of safety (Eicher and Kawachi 2011).

The city of the new collective life and collective spirit may well be growing within the localized boundaries of relatively empowered residents of high quality neighbourhoods. One of the traps of this recognition, however, is that planning for such neighbourhoods may in fact reduce the collective life potential in other parts of the city (DeFillippis et al. 2010). Some urban critics have shown that sustainability efforts which positively affect parts of the city may increase social injustices elsewhere (Burton 2000; Smith 2002; Quastel et al. 2012). Dooling (2009) devised the term "eco-gentrification" from her study of how the "environmental rationality and ethic" used to improve green space for housed residents in Seattle displaced and excluded the homeless – findings that contest the notion that green city policies are beneficial for all residents. Eco-gentrification as a phenomenon is not entirely new, but is specific to the increase of green city projects, as it "describes the convergence of urban redevelopment, ecologically-minded initiatives and environmental justice activism in an era of advanced capitalism" (Checker 2011: 212).

This can lead to the conclusion that "in fact sustainability and justice are like two rival brothers" (Mancebo 2015: 21). Referring to the case of revitalization of the social housing-dominated neighbourhoods in Paris, Mancebo reflects that a "good environment" by new urban design standards can lead to better living conditions, or the opposite, and the residents, their experiences, history, and expectations, are a big part of the determination of which way the pendulum will swing.

We cannot then expect a unified kind of community life enhancement strategy to apply in all cities. At the same time, neither can we expect enhanced community life to be equally beneficial to all groups of people. This is an additional aspect of the community trap: ignorance of the need for each community, made up of individuals with their own identities and aspirations, to determine its own sense of its social identity, what it is willing to share, and how. When we look at the suite of initiatives typically included within the realm of collective life in today's cities, from bikes to books, shared kitchens and gardens, cars and "meet-ups," there are certain topics of identity and class that are rarely if ever part of the package. This means that there are people who will rarely or never be part of the idealized collective. Advocates for those systematically excluded from urban collective life remind us that, as often as the question of equal rights and non-discrimination against people based upon their gender, sexuality, race, indigeneity or immigration status has been voiced, it has not been settled. As a result: "Many people in this cohort do not have confidence in their right to ownership, inclusion and belonging to the public spaces of the city because of the frequent reminders expressed by those who presume to judge and challenge those rights" (Griffin 2015: 8).

We can be more specific. The sense of collective spirit in community serves to further divide groups more than it unites them. Lokko (2015) describes the city of Johannesburg as quintessentially representing this phenomenon. Johannesburg is, to Lokko (ibid.: 12):

> at once a city of *anti*-collectives and *hyper*-collectives; endless satellites of tightly-knit, tightly-policed enclaves that sit uneasily together, bound by a network of freeways, roads, taxi routes and railway lines ... Within these enclaves, an exaggerated sense of community persists, an "*us vs. them*" attitude where the terms are interchangeable – one man's "us" is another's "them," and so on.

Breaking into these enclaves with genuine racial mixing, as has happened in some Johannesburg neighbourhoods, can often still bring disproportionate benefits to non-Black people and disproportionate risks to Black people. Talking about the case of American cities like Brooklyn, Moore (2015) makes the case that gentrification in historically Black neighbourhoods can be shown to put the lives of Black residents who remain there at greater risk, even as crime rates may decline. This is because of the persistent according of greater significance to white lives over Black lives: "Safety becomes a relative experience when gentrification occurs. The presence of white people almost always guarantees the increased presence of resources, like police, which does not always guarantee safety for black people in those same spaces" ( ibid.: 19–20). Community work, as Jeffrey et al. (2012: 1254) put it, "has its own walls" and exclusionary practices.

Even if we can adequately value all of our urban diversity, and the various commoning strategies that work within this diversity, there remains the question of the use to which we will put it. Miranda Joseph (2002) rejects the notion that work to build the public sphere and appreciate its diversity holds emancipatory potential, a third aspect of the community trap. Instead, according to Joseph, community work reinforces the norm of capitalist relations because doing community work is increasingly seen as dependent on "regular" work within capitalism to enable the leisure to do it. She describes the romantic narrative of 'community' as a place for local, face-to-face relations contrasted with the global and faceless logic of capitalism. Because of the structural contrast between this notion of community and modern capitalist society, she argues, community enables capitalism. She proposes that "precisely through being cast as its opposite, community functions in complicity with 'society,' enabling capitalism" (ibid.: 148). Community work and community life within the urban celebration exist alongside the capitalist system. This work therefore denies the energy and potential of work that might otherwise be done by communities acting outside the market, that might otherwise be creating a new social and experiential repertoire from which people may draw.

Joseph borrows from David Harvey and J.K. Gibson-Graham in support of this case for the inevitability of an increasingly globalized capitalist system. She points out that what on the surface are trends toward increasing attention to localism and community are not actually signs of the decline of mass production within global capitalism, but just the rise of increased product differentiation. In this drive for differentiated local places and products, localities and non-profit organizations take up a new form of service to global capital. They do this by pointing out the gaps into which for-profit production can move next, in the unquestioned and ongoing process of market differentiation and competition. Joseph, along with other critics of neoliberalism, argues for the supplementary role that community-based organizations play in expanding the reach of global capitalism by indicating the frontier that capitalist production has yet to cross. This is the non-profit and social work that others describe as "community-building." Joseph describes this – in Gramscian (Gramsci 1971) terms – as a "hegemonic apparatus, articulating the desire for community with a desire for capitalism" (Joseph 2002: 73).

## Conclusion

Passion for the city is back. The majority of humanity, and the mainstream of ideas about social progress, are now home in the city in a different way than was the case when Beauregard had cause to consider himself iconoclastic for choosing to live in Brooklyn. Still, Beauregard would not be satisfied with the current state of urban affairs (I asked him[4]). It turns out, two decades after that warning of the implications of losing passion for the city, we have to face the fact that while urban thinking may well be a precondition for moral thinking, moral thinking does not necessarily follow from urban thinking. Critical urban scholarship has maintained a sober view on the cognitive traps or errors possible in thinking about the potential to craft new political solutions based on the specific value of the local scale, engagement, and community work. The city is not a sure thing for seekers of an authentic life, an empowered life, or one with a social sense of the value of risk and the meaning of resilience. Nor does the rediscovery of the city as human habitat of choice ensure a right to live in the city for middle and lower classes.

Urbanists the world over are heartened by the contemporary celebration of the city and the prospect of cities as sites of social transformation. Much of this celebration is inadequately argued and justified, and each and every aspect which gives us cause for celebration in today's cities at least equally gives us cause for despair. At the same time, as a turn against the ravages of modernism in urban planning and design, against the brutality of exclusionary urban regime politics, and the isolation of the privatization agenda, the new urban celebration has much to recommend it to progressive urbanists. It just does not add up to the aspiration for cities that approach sustainability and justice. Better results for sustainable and just cities demand recognition of the many, diverse, valuable ways in which people may engage in the neighbourhood domain. The neighbourhood is a window into many other worldly domains, including both human and nonhuman communities, and serves diverse individual and social needs, in a range of forms and orientations of membership and attachment. As a set of complex social processes, habits, and beliefs, urban qualities and boundaries, empowerment and participation in decision-making, and community-building require careful analysis, rather than either romantic attachment or cynical rejection. Pragmatic efforts to advance our thinking and our actions in the urban domain can better reveal the opportunities to work for justice and sustainability in our contemporary cities.

Consideration of the call to the city that urbanists hear all around, and the cautions and warnings from astute critical urban scholars, lead us to the pragmatic, contemporary urban frontier. At this frontier, we can see plainly that the transformation being promoted and produced in cities around the world does not add up to the kind and the scale of transformation needed to produce sustainability and justice. If it is the city we want, and sustainability and justice we need, we need to look for approaches to a deeper transformation, and reasons to hope that these may be more possible today than ever before. It is to this task that we turn in the remainder of this book.

## Notes

1 Baird disagrees with this sentiment, while perceiving it to be dominant among most writers on design and architecture.
2 Beauregard also gives Davis a nod for his *City of Quartz* which appeared in 1990.
3 According to James:
    Grant an idea or belief to be true ... what concrete difference will its being true make in any one's actual life? What experiences [may] be different from those which would obtain if the belief were false? How will the truth be realized? What, in short, is the truth's cash-value in experiential terms? (James 1977 [1907]: 311)
4 Specifically, he said: "I do think there is a need for a more critical perspective on the current 'celebration' of the city ... [in 'Why passion for the city has been lost'] I was less interested in whether the city was 'celebrated' than whether it was hotly debated. All this current agreement does not meet my criterion." (RAB, 22 Jan. 2015, pers. comm.).

## References

Agrawal, Arun. "Common Resources and Institutional Sustainability." In *The Drama of the Commons*, edited by E. Ostrom. Washington, DC: National Academy Press, 2002, pp. 41–86.
Amin, Ash. (ed.) *Post-Fordism: A Reader*. New York: Wiley, 1994.
Baird, George. *Writings on Architecture and the City*. London: Artifice, 2015.
Banfield, Edward. *The Unheavenly City*. Boston: Little, Brown, 1968.
Barber, Benjamin. *If Mayors Ruled the World: Dysfunctional Nations, Rising Cities*. New Haven, CT: Yale University Press, 2013.

Barton, Hugh, Marcus Grant, Claire Mitcham, and Catherine Tsourou. "Healthy Urban Planning in European Cities." *Health Promotion International* 24 (2009): i91–i99.

Beauregard, Robert A. "Why Passion for the City Has Been Lost." *Journal of Urban Affairs* 18(3) (1996): 217–231.

BIG (Bjarke Ingels Group). *Yes is More.* Copenhagen: Taschen Books, 2016.

Brenner, Neil and Christian Schmid. "The 'Urban Age' in Question." *International Journal of Urban and Regional Research* 38(3) (2014): 731–755.

Brenner, Neil and Christian Schmid. "Towards a New Epistemology of the Urban?" *City* 19(2–3) (2015): 151–182.

Brugmann, Jeb. *Welcome to the Urban Revolution: How Cities Are Changing the World.* New York: Bloomsbury Press, 2009.

Bulkeley, Harriet. "Urban Sustainability: Learning from Best Practices?" *Environment and Planning A* 38 (2006): 1029–1044.

Burton, Elizabeth. "The Compact City: Just or Just Compact? A Preliminary Analysis." *Urban Studies* 37(11) (2000): 1969–2006.

C40 Cities. 2011 (Dec. 23) "Case Study: Climate Action in Major Cities, C40 Cities Baseline and Opportunities." Available at: www.c40.org/case_studies/climate-actio n-in-major-cities-c40-cities-baseline-and-opportunities (accessed 18 August 2016).

Cannuscio, Carolyn and Karen Glanz. "Food Environments." In *Making Healthy Places: Designing and Building for Health, Well-being, and Sustainability*, edited by Andrew Dannenberg, Howard Frumkin, and Richard Jackson. Washington, DC: Island Press, 2011, pp. 50–62.

Capon, Anthony. "The Way We Live in Our Cities." *Medical Journal of Australia* 187 (2007): 658–661.

Castel, Robert. *Metamorphose de la question sociale: une chronique du salariat.* Paris: Fayard, 1995.

Castells, Manuel. *The Urban Question.* Cambridge, MA: MIT Press, 1977.

Castells, Manuel. *The City and the Grassroots.* Berkeley, CA: University of California Press, 1983.

Checker, Melissa. "Wiped Out by the 'Greenwave': Environmental Gentrification and the Paradoxical Politics of Urban Sustainability." *City & Society* 23(2) (2011): 210–229.

City of Portland. "Public Involvement Principles." Portland, OR: City of Portland, 2010.

Congrès internationaux d'architecture moderne (CIAM), *La Charte d'Athènes or The Athens Charter.* Trans. J. Tyrwhitt. Paris: The Library of the Graduate School of Design, Harvard University, 1946 [1933].

Cruikshank, Barbara. *The Will to Empower: Democratic Citizens and Other Subjects.* Ithaca, NY: Cornell University Press, 1999.

Cruz, Teddy and Fonna Forman. "Public Imagination, Citizenship and an Urgent Call for Justice." In *The Just City Essays: 26 Visions for Urban Equity, Inclusion and Opportunity*, edited by Toni Griffin, Ariella Cohen, and David Maddox. V.1. New York: J. Max Bond Center on Design, Spitzer School of Architecture, City College of New York, Next City andThe Nature of Cities, 2015, pp. 40–45.

Dannenberg, Andrew, Howard Frumkin, and Richard Jackson (eds.) *Making Healthy Places: Designing and Building for Health, Well-being, and Sustainability.* Washington, DC: Island Press, 2011.

Das, P.K. "Claiming Participation in Urban Planning as a Right." In *The Just City Essays: 26 Visions for Urban Equity, Inclusion and Opportunity*, edited by Toni Griffin, Ariella Cohen, and David Maddox. V.1. New York: J. Max Bond Center on Design, Spitzer School of Architecture, City College of New York, Next City and The Nature of Cities, 2015, pp. 91–95.

Davis, Mike. *City of Quartz*. London: Verso, 1990.

DeFillippis, James, Robert Fisher and Eric Shragge. *Contesting Community: The Limits and Potential of Local Organizing*. New Brunswick, NJ: Rutgers University Press, 2010.

De Shalit, Avner and Daniel Bell. *The Spirit of Cities: Why the Identity of a City Matters in a Global Age*. Princeton, NJ: Princeton University Press, 2011.

Dooling, Sarah. "Ecological Gentrification: A Research Agenda Exploring Justice in the City." *International Journal of Urban and Regional Research* 33(3) (2009): 621–639.

Dryzek, John and Simon Niemeyer. *Foundations and Frontiers of Deliberative Governance*. Oxford: Oxford University Press, 2010.

Ehrenhalt, Alan. *The Great Inversion and the Future of the American City*. New York: Alfred A. Knopf, 2012.

Eicher, Caitlin and Ichiro Kawachi. "Social Capital and Community Design." In *Making Healthy Places: Designing and Building for Health, Well-being, and Sustainability*, edited by Andrew Dannenberg, Howard Frumkin, and Richard Jackson. Washington, DC: Island Press, 2011.

Ewing, Reid and Robert Cervero. "Travel and the Built Environment: A Meta-analysis." *Journal of the American Planning Association* 76(2010), 265–294.

Fitzgerald, Joan. *Emerald Cities: Urban Sustainability and Economic Development*. New York: Oxford University Press, 2010.

Florida, Richard. *The Rise of the Creative Class … and How It's Transforming Work, Leisure, Community and Everyday Life*. New York: Basic Books, 2002.

Francis, Jacinta, Lisa Wood, Matthew Knuiman, and Billie Giles-Corti. "Quality or Quantity? Exploring the Relationship Between Public Open Space Attributes and Mental Health in Perth, Western Australia." *Social Science and Medicine* 74(2012): 1570–1577.

Freytag, Tim, Stefan Gossling, and Samuel Mossner. "Living the Green City: Freiburg's Solarsiedlung Between Narratives and Practices of Sustainable Urban Development." *Local Environment* 19(6) (2014): 644–659.

Frumkin, Harold. "Urban Sprawl and Public Health." *Public Health Reports* 117(2002): 201–217.

Fullilove, Mindy T. "An Antidote for the Unjust City: Planning to Stay." In *The Just City Essays: 26 Visions for Urban Equity, Inclusion and Opportunity*, edited by Toni Griffin, Ariella Cohen, and David Maddox. V.1. New York: J. Max Bond Center on Design, Spitzer School of Architecture, City College of New York, Next City and The Nature of Cities, 2015, pp. 82–84.

Fung, Archon. "Recipes for Public Spheres: Eight Institutional Design Choices and Their Consequences." *The Journal of Political Philosophy* 11(3) (2003): 338–367.

Gebel, Klaus, Adrian Bauman, Neville Owen, Sarah Foster, and Billie Giles-Corti. "Position Statement: The Built Environment and Walking." Sydney: National Heart Foundation of Australia, 2009.

Giles-Corti, Billie, Kate Ryan, and Sarah Foster. "Increasing Density in Australia: Maximising the Benefits and Minimising the Harm." Sydney: National Heart Foundation of Australia, 2012.

Glaeser, Ed. *The Triumph of the City*. New York: Penguin, 2011.

Gleeson, Brendan. "Critical Commentary: The Urban Age: Paradox and Prospect." *Urban Studies* 49(5) (2012): 931–943.

Graham, Stephen and Simon Marvin. *Splintering Urbanism: Networked Infrastructures, Technological Mobilities and the Urban Condition*. New York: Routledge, 2001.

Gramsci, Antonio. *Selections from the Prison Notebooks*. New York: International Publishers, 1971.

Griffin, Toni. "Defining the Just City Beyond Black and White." In *The Just City Essays: 26 Visions for Urban Equity, Inclusion and Opportunity*, edited by Toni Griffin, Ariella Cohen, and David Maddox. V.1. New York: J. Max Bond Center on Design, Spitzer School of Architecture, City College of New York, Next City and The Nature of Cities, 2015, pp. 6–9.

Groth, John, and Eric Corijn. "Reclaiming Urbanity: Indeterminate Spaces, Informal Actors and Urban Agenda Setting." *Urban Studies* 42(3) (2005): 503–526.

Hartig, Terry, Gary Evans, Larry Jamner, Deborah Davis, and Tommy Gärling. "Tracking Restoration in Natural and Urban Field Settings." *Journal of Environmental Psychology* 23 (2003), 109–123.

Healey, Patsy. *Urban Complexity and Spatial Strategies: Towards a Relational Planning for Our Times*. New York: Routledge, 2007.

Healey, Patsy. "In Search of the 'Strategic' in Spatial Strategy Making." *Planning Theory & Practice* 10(4) (2009): 439–457.

Hertin, Julia and Frans Berkhout. "Analyzing Institutional Strategies for Environmental Policy Integration: The Case of EU Enterprise Policy." *Journal of Environmental Policy & Planning* 5(1) (2003): 39–56.

Holden, Meg. "Public Participation and Local Sustainability: Questioning a Common Agenda in Urban Governance." *International Journal of Urban and Regional Research* 35 (2) (2011): 312–329.

Jacobs, Allan B. *The Good City: Reflections and Imaginations*. New York: Routledge, 2011.

James, William. "Pragmatism and Radical Empiricism." In *The Writings of William James*, edited by J.J. McDermott. Chicago: University of Chicago Press, 1977 [1907], pp. 311–317.

Jeffrey, Alex, Colin McFarlane, and Alex Vasudevan. "Rethinking Enclosure: Space, Subjectivity and the Commons." *Antipode* 44(4) (2012): 1247–1267.

Joseph, Miranda. *Against the Romance of Community*. Minneapolis, MN: University of Minnesota Press, 2002.

Kanra, Basra. "Binary Deliberation: The Role of Social Learning and the Theory and Practice of Deliberative Democracy." Joint Sessions. Helsinki: European Consortium for Political Research, 2007.

Katz, Bruce. "Report from the Metropolitan Revolution." Brookings Blog. Available at: www.brookings.edu/blog/the-avenue/2014/12/11/report-from-the-metropolita n-revolution/ 2014 (11 Dec.) (accessed 2 Nov. 2016).

Katz, Bruce and Jennifer Bradley. *The Metropolitan Revolution*. Washington, DC: Brookings Institution Press, 2013.

Kennedy, Christopher. *The Evolution of Great World Cities*. Toronto: University of Toronto Press, 2011.

Kent, Jennifer, Susan Thompson, and Anthony Capon. "Healthy Planning." In *Planning Australia: An Overview of Urban and Regional Planning*, edited by Susan Thompson and Paul Maginn, 2nd edn. Port Melbourne, Victoria: Cambridge University Press, 2012.

Kipfer, Stefan. "Why the Urban Question Still Matters: Reflections on Rescaling and the Promise of the Urban." In *Leviathan Undone? Towards a Political Economy of Scale*, edited by Roger Keil and Rianne Mahon. Vancouver: UBC Press, 2009, pp. 67–83.

Koohsari, Mohammad, *et al.* "Public Open Space, Physical Activity, Urban Design and Public Health: Concepts, Methods and Research Agenda." *Health & Place* 33 (2015): 75–82.

Lokko, Lesley. "In It Together." In *The Just City Essays: 26 Visions for Urban Equity, Inclusion and Opportunity*, edited by Toni Griffin, Ariella Cohen, and David Maddox. V.1. New York: J. Max Bond Center on Design, Spitzer School of Architecture, City College of New York, Next City and The Nature of Cities, 2015, pp. 10–14.

Lowe, Melanie and Billie Giles-Corti. "Cities and Health: Preventing NCDs Through Urban Design." In *Dancing in the Rain*, edited by Grant Blashki and Helen Sykes. Albert Park, Victoria: Future Leaders, 2015.

Magnusson, Warren. *The Politics of Urbanism: Seeing Like a City*. New York: Routledge, 2013.

Mancebo, François. "Ceci n'est pas une pipe: Unpacking Injustice in Paris." In *The Just City Essays: 26 Visions for Urban Equity, Inclusion and Opportunity*, edited by Toni Griffin, Ariella Cohen, and David Maddox. V.1. New York: J. Max Bond Center on Design, Spitzer School of Architecture, City College of New York, Next City and The Nature of Cities, 2015, pp. 21–25.

Marshall, Alex. *How Cities Work: Suburbs, Sprawl, and the Roads Not Taken*. Austin, TX: University of Texas Press, 2001.

Marshall, Alex. *The Surprising Design of Market Economies*. Austin, TX: University of Texas Press, 2013.

Merrifield, Andy. *Dialectical Urbanism: Social Struggles and the Capitalist City*. New York: Monthly Review Press, 2002.

Merrifield, Andy. *The New Urban Question*. London: Pluto Press, 2014.

Moore, Darnell. "Urban Spaces and the Mattering of Black Lives." In *The Just City Essays: 26 Visions for Urban Equity, Inclusion and Opportunity*, edited by Toni Griffin, Ariella Cohen, and David Maddox. V.1. New York: J. Max Bond Center on Design, Spitzer School of Architecture, City College of New York, Next City and The Nature of Cities, 2015, pp. 18–20.

Project for Public Spaces. "What is Placemaking?" 2015. Available at: www.pps.org/reference/what_is_placemaking/ (accessed 18 August 2016).

Purcell, Mark. "Urban Democracy and the Local Trap." *Urban Studies* 43(11) (2006): 1921–1941.

Quastel, N., M. Moos and N. Lynch. "Sustainability-as-Density and the Return of the Social: The Case of Vancouver, British Columbia." *Urban Geography* 33(7) (2012): 1055–1084.

Rao, Mala, Sunand Prasad, Fiona Adshead, and Hasitha Tissera. "The Built Environment and Health." *Lancet* 307(2007): 1111–1113.

Rose, Jonathan. *The Well-Tempered City*. New York: Harper Collins, 2016.

Ross, Catherine and John Mirowsky. "Neighborhood Disadvantage, Disorder, and Health." *Journal of Health and Social Behavior* 42(2011): 258–276.

Rybczynski, Witold. *Makeshift Metropolis: Ideas About Cities*. New York: Scribner, 2010.

Saelens, B. E., et al. "Neighborhood Environment and Psychosocial Correlates of Adults' Physical Activity." *Medicine and Science in Sports and Exercise* 44: 637–646.

Sallis, James, Rachel Millstein, and Jordan Carlson. "Community Design for Physical Activity." In *Making Healthy Places: Designing and Building for Health, Well-Being, and Sustainability*, edited by Andrew Dannenberg, Howard Frumkin, and Richard Jackson. Washington, DC: Island Press, 2011, pp. 33–49.

Sandercock, Leonie. *Towards Cosmopolis: Planning for Multicultural Cities*. London: John Wiley & Sons, Ltd, 1998.

Sassen, Saskia. *The Global City*. Princeton, NJ: Princeton University Press, 1991.

Schumacher, E. Fritz. *Small is Beautiful: Economics as if People Mattered*. London: Blond & Briggs, 1973.

Scott, James C. *Seeing Like a State: How Certain Schemes to Improve the Human Condition Have Failed*. New Haven, CT: Yale University Press, 1998.

Smith, Neil. "New Globalism, New Urbanism: Gentrification as Global Urban Strategy." *Antipode* 34(2002): 427–450.

Speck, Jeff. *Walkable City: How Downtown Can Save America, One Step at a Time.* New York: Farrar, Strauss and Giroux, 2012.

Stafford, Mai, Tarani Chandola, and Michael Marmot. "Association Between Fear of Crime and Mental Health and Physical Functioning." *American Journal of Public Health* 97(2007): 2076–2081.

Sussmann, Cornelia. "Toward the Sustainable City: Vancouver's Southeast False Creek." PhD dissertation. University of British Columbia, 2012.

Swartling, Asa G., Mans Nilsson, Rebecca Engstrom, and Lovisa Hagborg. "Theory and Methods for EPI Analysis." In *Environmental Policy Integration in Practice: Shaping Institutions for Learning,* edited by Mans Nilsson and Katarina Eckerberg. London: Earthscan, 2007, pp. 49–66.

The Healthy Built Environments Program. *Healthy Built Environments: A Review of the Literature. Fact Sheets.* Sydney: Cities Futures Research Centre, The University of New South Wales, 2012.

Thévenot, Laurent. "Voicing Concern and Difference: From Public Space to Common Places." *European Journal of Cultural and Political Sociology* 1(1) (2014): 7–34.

United Nations Human Settlements Programme. *The Challenge of Slums: Global Report on Human Settlements.* London: Earthscan, 2003.

Westerhoff, Lisa. "City Stories: From Narrative to Practice in Vancouver's Olympic Village." PhD dissertation. University of British Columbia, 2015.

World Health Organization. *Global Status Report on Non-Communicable Diseases 2010.* Geneva: World Health Organization, 2011.

Young, Iris M. "Communication and the Other: Beyond Deliberative Democracy." In *Democracy and Difference: Contesting the Boundaries of the Political,* edited by Seyla Benhabib. Princeton, NJ: Princeton University Press, 1996, pp. 120–136. Zarate, Loreno. "Right to the City for All: A Manifesto for Social Justice in an Urban Century." In *The Just City Essays: 26 Visions for Urban Equity, Inclusion and Opportunity,* edited by Toni Griffin, Ariella Cohen, and David Maddox. V.1. New York: J. Max Bond Center on Design, Spitzer School of Architecture, City College of New York, Next City and The Nature of Cities, 2015, pp. 26–29.

# Part II

# An urban way forward in a pragmatic view

The remainder of this book describes "what gives" in the contemporary urban celebration, this sea-change in thinking about the city, as an undercurrent of ethical urban promise and a path toward justice and sustainability. First, though, a warning: this promise of a future-moral engagement is mostly unspoken, certainly not dominant, and at any rate not any kind of ideal that a designer of a better city would choose *prima facie*. Instead, it represents three pragmatic access points by which the city, as urbanites may experience it today, makes its bid for our passions in a way that frames a move toward common moral action, an urban common place. It is a response to each of the local trap, the empowerment trap, and the community trap presented by critical urban scholars to temper the present urban celebration. The corresponding perspective that we will examine is: the value within the new, urban notion of authenticity (Chapter 4); the potential for specifically urban empowerment (Chapter 5); and how the city offers a resilience-based stance toward the risk in community action that is conducive to valuing human diversity (Chapter 6). Taken together, these possibilities constitute a uniquely urban, pragmatic approach to sustainability and justice.

In Chapter 4, we examine the implications of a new urbane understanding of authenticity, its challenges and opportunities for justice based upon the pragmatic understanding of the personal regime of engagement that dominates under conditions where authenticity is key. Important dynamics of the authenticity regime to be drawn out for their implications in seeking sustainability and justice include the public realm in new urban environments and notions of open-minded space in the city. In Chapter 5, we will question the potential in cities today for a new variety of empowerment, explicitly anti-utopian, sometimes even anti-political. We will examine the way in which a pragmatic understanding of the dynamics of argumentation in the public sphere offers a new understanding of empowered urban citizenship, in which the work being undertaken in some urban communities today could be revalued. In Chapter 6, we will take a look at the pragmatic understanding of the discursive, democratic scene of struggle, and in particular the regimes of engagement by which arguments in the public sphere offer a way of understanding how urbanites value risk and safety, and steps that may be underway toward a more socially resilient urbanism. In each case, we will seek evidence of the way in which our contemporary cultural condition of urbanity emits evidence of possible shifts and reconceptualizations, and how a pragmatic understanding can help reveal these shifts.

This construct of uniquely urban qualities of life emerges from a personal and experiential perspective, as I feel the potential of these urban qualities within the city's grip. At the same time, I am not the first to single out this nexus of urban traits as remarkable and worthy of cultivation. Beauregard and Bounds (2000), in describing what a specifically urban citizenship requires and offers, point out five key clusters of rights and responsibilities as central to urban citizenship. The first two are safety and tolerance, comparable to the risk-reward nexus that I will discuss in Chapter 6. Third is political engagement, which is in keeping with the urban sense of empowerment discussed in Chapter 5. Fourth and fifth are recognition and freedom, aligned with the authenticity trait singled out in Chapter 4.

The additional reason for honing in on these elements of authenticity, empowerment, community and risk is that each has been subject to considerable critique within the urban literature. That is, as discussed in Chapter 3, the discomfort of critical scholars with each of these concepts has given rise to critique of, respectively, the local trap, the empowerment trap, and the community trap. This critical attention to the city, which exists quite separately from the urban celebration literature and in much more rarefied locales of urban scholarship, emerges out of a concern for the fate of justice and democracy under conditions of neoliberal, globalized capitalism.

Sustainability, of course, as a potentially emancipatory and integrative concept, is also at risk of remaining in a stunted phase of implementation to the extent that it is not connected explicitly and continuously to political processes of open discourse. Political theorists Dryzek and Niemeyer (2010: 128) explain how limiting public dialogue around sustainability can limit the breadth of the approach taken:

> Sustainable development rests on the ultimate compatibility and mutually reinforcing interaction of economic growth, environmental conservation and social justice. Its history shows it being moved in an ever more business-friendly direction. This means that a range of critical questions get sidelined. These include the ultimate compatibility of capitalist economic growth with ecological limits and the interests of non-human nature and ecosystemic integrity of the sort that radical greens are likely to recognize. Now, defenders of sustainable development might well argue that all these critical questions can be addressed within the terms of what can be a fairly flexible discourse. But the sustainable development discourse is not infinitely elastic, and the concerns just listed lie outside its limits.

These macro-scale trends of neoliberalism, globalization, privatization, and securitization, as they narrow the political options available to cities and other democratic movements and groups, are the focus of these critical scholars' concern for the fate of democratic values and actions in the city. The pragmatic path forward offered here responds to these critiques, without being petrified by them. I attempt here to piece together the argument that we are witness to new deconstructions of injustice, and offer possible new constructions of what constitutes the common good, in the city today.

In no way do I intend to suggest in the following three chapters that urban life must represent the characteristics of authenticity, empowerment, and risk resilience. I am not even on solid ground in suggesting that they *usually* represent them. The urban scale is now and has always been used in the opposite way for the purposes of limiting human hope and freedom, disempowering people and groups, inducing fear and submission to the will of the powerful and the demands of capital in particular. What I would like to argue, by contrast, is that urban life still retains the prospect of emancipatory, justice- and solidarity-seeking ends – the same ends that are also requisite of a move toward a sustainable society. What is more remarkable is that contemporary urbanity may retain some *unique* capacity to move toward these ends. If our notions of the contemporary city are to merit their current celebration, they must give us reason to view the city as a common place in which diverse people find moral value in seeking sustainability and justice. It is thus in a spirit of deep-city exploration that I offer these possibly unique, and distinctly moral, urban values. I offer it along with the challenge that if this construct is right, then urbanists ought to be more engaged in studying and promoting these values, if we want cities that live up to the celebratory tone that is being used on them at present. In the second half of this book, we seek to put the urban celebration to a kind of reality test by which we will be able to determine the pragmatic "cash value" of certain dominant urban trends. By pragmatic "cash value," I mean a measure of the way in which contemporary urbanists' moral intuitions about the city translate into practices in the diverse lives being lived in cities today.

The structure of Chapters 4, 5, and 6 is designed to introduce a key critical resource from the corpus of pragmatism, a tool for thinking deeply about the meaning and potential of radical liberal democracy. This tool for thinking and action will be sunk into the ground of our present challenge of building cities that are better suited to addressing sustainability and justice goals with the help of commentary and insight from other theorists and activists. Our challenge.

Allow me to add a note about my use of "our" here, as well as "we" in spots, in what follows. Is the use of such assuming pronouns not evidence of my slipping into the fold of universal assumptions that ought to be anathema to a pragmatic approach? The answer is: yes, but let me explain. Yes, the use of "we" with you, my readers of untold diversity of thought and potential (certainly a diversity and potential that is untold to *me*, because I do not know you well enough to know such things about you), lays a claim upon your thinking that is unearned. I offer the "we" in experimental format, as an invitation to try my thinking about this urban moment and its potential, on for size. On its own, the use of "we" would reduce the effort of the second half of this book to bare naked proselytization. That is not what I am up to. I will offer you experiential and studied empirical details of how I have arrived at my understanding so far. I will offer quotations and citations from others, whom you may be more inclined to believe and whom I will interpret as seeing the city and its opportunity structure in this way, too. For some of you, this conventional social scientific ballast may suffice to make good on the claim to include you as part of the "we" in my argument here, and to incorporate these ideas into your own way of thinking about the paths to justice and sustainability from our urbane present.

For others of you, I know it will not. You may cling to an understanding of a preferable past to which some of us might return, whether characterized by an idealized (or at least more universal) welfare state, a stronger compulsion and competence for local self-sufficiency, a more effective cultural tuning to the more-than-human world, or some other ideal. Your own experiences and experiments may have led you to a level of skepticism about what I see as the potential in urbanity for justice and sustainability that seems beyond return. For you, it is against your will that I include you in my "we." It is, in part, a plea. It is also an invitation. Because my experimental aim here is to sketch the contours of an urbanity that all of us do share, and offer space for sustainability and justice visioning within that sketch. I believe this aim is within and between all of us; and that all of us have different degrees of alienation from this experiment, as well. Because it is up to all of us to take up the discursive battle, even though there is no ultimate end point of agreement and unity in my view of this experiment, that is no excuse to shy away from the "we." It is, in short, an utterly pragmatic "we."

Not all of those whose thoughts will be used to build the argument here consider themselves pragmatists; but most see themselves as urbanists, with varying levels of hope for the city in the realm of sustainability and justice. The objective of our argument in each chapter is to justify a pragmatic process of identifying how to better reflect and operationalize the urban sustainability and justice potential. Ideas, critiques, cases, and images of the city from a range of different perspectives will be presented as a diverse array of evidence for understanding the city as holding unique moral potential within the drive for authenticity, empowerment, and risk resilience. I do not expect all of my readers to be persuaded by the full gamut. Not all urbanites are driven by the same motivations, and not all recognize the same set of urban values. This is a necessary and a great fact about cities and their people. My task instead is to open enough windows that there will be one fit for nearly everyone to climb through, into a common world of both habits and structures that will promote more action toward sustainability and justice.

## References

Beauregard, Robert and Anna Bounds. "Urban Citizenship." In *Democracy, Citizenship and the Global City*, edited by Engin Isin. New York: Routledge, 2000, pp. 243–256.
Dryzek, John and Simon Niemeyer. *Foundations and Frontiers of Deliberative Governance*. Oxford: Oxford University Press, 2010.

# 4 An urban shot at authenticity
## Our cities become ourselves

The appeal of authenticity is strong in today's cities. In this chapter, we examine the implications of authenticity as a widely embraced urban ideal. From the perspective of the pragmatic sociology of engagement, the individualist emphasis on authenticity eludes a traditional collectivist justification for public debates, and thus stifles attempts to revalue justice and sustainability in community. Urban critics from outside the pragmatic tradition have noted the shortcomings of an excessive emphasis upon individual authenticity too. We can thus recognize the dangers of urban authenticity. Many critics stop here. But a pragmatic understanding of the dynamics of engagement in the public realm offers a different interpretation of the role that individualist authenticity can play in public. We will examine this route to understanding the role of urban authenticity, and the particular value of multiplicity in urban identities in the work to seek compromises in public disputes. Through reaching compromises, authentic individuals in the city can find potential new paths to connect without necessarily sacrificing their deeply held values. This possibility exists because they have the opportunity, when acting in an urban mode, to see their authentic identity as changeable and dynamic, more like a "personality" they can adopt than a hard-wired essence. This makes compromises more likely to arise, and also suggests the potential emergence of a justice element within compromises, by which those who engage come to envisage urban change but actually recognize "a right to change ourselves by changing the city" (Harvey 2008: 23). This chapter seeks analytical ways to understand the process of putting habits of living authentic, urban human lives to more justice-serving pursuits.

## The implications of individualism for urban notions of justice

A pragmatic understanding of the possibilities for justice emphasizes the changeable nature of justice for people who engage in public argument that seeks to attain it. In this view, people's behaviours are fundamentally uncertain and over-determined by a plurality of causal forces. People also regularly mobilize scenarios and arguments that are analytically represented by distinctive models of justification and regimes of engagement which demand, one assumes, a relational position. These arguments are changeable, and regularly conflict. People must be willing to recognize the distance between the way they view the path to justice, and what other people see, and be willing to adjust their stance with a view to compatible outcomes, if consent for a single action in view of justice is to be gained. In many cases, even the most mundane, this resolution of the tension in between views of justice is not at all obvious: "Since the principles of justice invoked are not immediately compatible, their presence in a single space leads to tensions that have to be resolved if the action is to take its normal course" (Boltanski and Thévenot 2006: 216).

In presenting his pragmatic sociology of engagement, Thévenot finds the most commonly operating regime of engagement to be one oriented around personal familiarity, as opposed to, say, class- or civic-based solidarity. This regime of engagement includes heavy doses of individualism and difference, that shade the nature of disputes and understandings of justice brought to light by urban spaces and city life today. There are losses in this type of engagement when it comes to the explicitly political content of dialogue and discourse. A pragmatic approach recognizes that the need for authenticity with self, separate from the social group, is a countervailing force that limits a culture's ability to form and maintain a community as a reference group of worth, of justice. And there are gains, too, in terms of what we can consider liberal freedoms: the freedom to define and assert the value of one's personalized life values, contexts, belongings, relationships, preferences. These freedoms become more legitimized in the personal and familiar regime of engagement when an individual engages in their own individual plan, and projects their will and their preferences into the future, as a goal to move towards with a sense of conviction for one's chosen habits, and intention to progress as an individual. In engaging in this kind of personal life planning, the individual asserts a sense of self-assurance and orientation toward the future. This re-values the practices they engage in today (Thévenot 2014). Through these practices, one becomes authentically one's self.

In the public domain, this regime of engagement results in a "multi-layered fabric of personalities" (ibid.: 13), some better and some less well defined, intentionalized, projected into the future. It also results in inner tensions which may cause them to change course drastically from time to time, offering new sets of reasons at each successive about-face. Such changes do not threaten the presumed authenticity of the individual. Rather, what does threaten it is associating one's individual freedom and individual life planning too closely with any group plan, ideal, or project. Thus, the quality of engagement in the civic realm, where one's individual freedom must sometimes take a back seat to the good of the populace as a whole, or to a notion of equality or redistribution, suffers when so much emphasis is placed on the familiar regime of engagement. This tension constitutes a key challenge in the construction of authenticity:

> This [familiar] regime [of engagement] is a prerequisite for notions of autonomy, individual responsibility, choice, project-making, contract ... It expanded at the expense of other commitments and associated evaluations,

such as solidarity and egalitarian commitment to a civic order of worth. It thus threatens the engagement in civic qualification that formerly prevailed.

(ibid.: 14)

The dynamic nature of the good within the familiar regime of engagement is one of its key features. This dynamism leads Thévenot to consider the term "personality" more appropriate than the term "individual" when referring to agents engaging in public dialogue within this regime. The term personality seems to point out the changeable and sometimes unpredictable values and habits that may be brought to the fore in such dialogues. An individual may represent a fair number of personalities, depending on context, timing, and associations. This regime opens up "a more dynamic approach to the notion of identity" (ibid.: 11) than the solitary, all-encompassing notion of identity than the term typically connotes. This pragmatic notion of identity as personality is neither entirely flexible and changeable nor entirely fixed and intransient. It is set in part by the material context being engaged, in part by the individual's sense of progress and change needed in crafting and effecting their own life plan. In recognition of this puzzle-piecing of identity fragments in the modern individual who engages in public and political discourse, we can consider the individual as: "a combination of disparate components that have to be subsumed under some kind of dynamic identity" (ibid.: 9).

The fact that this identity is dynamic, but not entirely chaotic, is based strongly upon the material context in which individuals craft their life plans and habits, which determines identity constructs to a greater degree than is often realized by the hypothetical liberal freedom-loving individualist. Thévenot refers to these material contexts as "formatted environments." The new urban design tricks, technologies, and preferences of placemaking may be considered as "investments in form" required to shape individual constructions of changeable identities in ways that are conducive to further preference-building along the same lines as the design changes themselves. In his conception, as the individual develops daily and adopts special habits of engaging in these kinds of environments, and as their individual life plan is affected by these daily habits and emergent preferences, the need for individuals to have a sense of "continuity of the self, from one place and time to another," constitutes a "pledge" to this material context. The pledge is a "guarantee, or security, deposited in a proper environment" (ibid.: 11) that over time constitutes an "equipped humanity" for making the most of that space for further social and public engagements (Thévenot 2002).

The new urban design orthodoxy which has emerged as part of the rediscovery of the city in the past 15 years plays well into this individualization of our notion of urban justice. Dense, compact and mixed-use neighbourhoods, with active sidewalks and pedestrian realms, including front stoops on low- and mid-rise residential buildings, are presented as ideal places in which urbanities can authentically *be*. Freed at one extreme from dependence on automobiles that subject suburbanites to separation from their home environments, and at the other from dependence on elevators, that subject urbanites in very high density neighbourhoods to radical isolation from one another, urbanites in these ideal placemade urban neighbourhoods are faced with an environment ideally suited for street-level engagement. "Eyes on the street," as Jane Jacobs evocatively put it (Jacobs 1961: 35). The small shops idealized in this kind of environment demand many interactions with shop keepers, not simply one interaction with a mechanized check-out system at a big box store. The pop-up markets, pocket parks, food trucks on the street, coffee counters fronting the sidewalk, and other casual spaces of interaction, also offer urbanites ample opportunities to engage with others in a certain individualized and authentic style.

Things can move quickly in the urban mode, often with great intensity. This tendency can favour accidental over strategic outcomes, can quickly transform local, everyday actions to actions with global-scale significance, and an impulse to act that comes from within individuals and groups rather than from some outside antagonism. When we see the difference that this urban mode of behaviour makes, the solution set for our problems changes. Warren Magnusson (2013: 53) offers this admiration of what we may get from seeing like a city:

> to see like a city is to claim the world as our own, rather than treat it as an alien production. It may be a frightening world, but we can begin to make sense of it if we work outward from the urban realities we know.

The new urban design is intended to welcome residents and visitors to spend more time in this sort of interaction with their local geography and social landscape, to associate their identity with it, and perhaps, to "become native to this place," to borrow a book title from Wes Jackson (1994). And so, residents of these kinds of urban environments become equipped with a sense of authenticity within just such a material and social context. Whereas Jackson would hate for his book title to be associated with an argument about urban environments, because he intended to evoke a need to reconnect our identity to local soils, plants, seasonality, and other patterns in nonhuman nature, the new urban village lays claims to an urban authenticity that can be just as fulfilling of a deep human need to belong.

Many people have responded to the appeal of this proposition to become native to their new urban neighbourhoods. This type of longing is compelling and enduring. At the beginning of the twentieth century, Patrick Geddes's (1970 [1905]) civic sociology generated the view that social and cultural phenomena must be seen in their regional context. Following Le Play, he argued that the physical environment to which people must adapt their actions provides a locale or habitat that constrains the possibilities of action that people can pursue. Their way of life, therefore, reflects the limits and opportunities that are inherent in the structure and conditions of the locale, understood as a spatially and socially defined region. Villages, towns, and cities develop within their surrounding region and Geddes held that sociology must necessarily investigate all social activities in relation to the region. Social actions can be freely pursued within these constraints, and cultural values and ideas are thus able to shape and transform the environment. There is, therefore, a two-way, reciprocal determination of environment and culture.

Fast forward to the present day, the new practices of "placemaking" also represent a revival of the concept of the public interest and of public space as a democratic demand. Building upon the political idea of serving common needs, placemaking in public space associates a political notion of the public interest with physical space attributes of non-excludability. Walter Lippmann, in the 1920s, may have offered the most popular definition of the public interest to date, as: "what men [*sic*] would choose if they saw clearly, thought rationally, acted disinterestedly and benevolently" (1922: 42). There is an independent but common construction to these arguments about placemaking and the public interest. For PPS, the public interest can be served as a synergistic outcome of good urban design. For Lippman, the public interest can be served as a synergistic outcome of good dialogue. In both cases, the politics of the public interest are best handled obliquely, by focusing on other things: good physical space attributes, and open-minded and informed speech conditions. From a pragmatic perspective, it will take more than either of these considerations to build and maintain strong communities that respect both justice and sustainability.

PPS focuses on the spatial notion of neighbourhood as a means to create and serve the public interest. This concept of better urban design as a means to advance authenticity in both individual urban lifestyles and physical urban spaces is being advanced with verve and avarice by cities worldwide. With attention to diversity and openness as well as the particular aesthetics of authenticity, this reconstruction will do its part to maintain and rebuild as much diversity as the city can accommodate. Harking back to Kevin Lynch (1960), but including others like Jane Jacobs (1961), Allan Jacobs (2011), and Christopher Kennedy (2011), authenticity through this design lens is tantamount to image-ability. That is, good design generates meaning- and sociability-inducing spaces, which become irresistible.

For all the renewed attention to placemaking, the public quality of public space cannot be reduced to questions of physical design alone. Places have political dimensions. Political dimensions demand direct, not just glancing, attention, when it is urban civic culture, the community, that needs cultivating in these spaces. It must be asserted too that not all urbanites consent to have their preferences, values, ideas, and discourses determined by just this kind of "placemade" environment. There is wide, cross-class, cross-culture appeal to key design features like pedestrian and other active pathways that wind in between buildings, offering a chance to witness and interact with diverse art forms from graffiti to public pianos to giant robotic sculptures, comestibles that come to you by truck or kiosk and offer gateways to different cultures, organized and pop-up experiences to learn something or meet someone new. These urban designs offer the prospect of being inspired by watching, in curated and more open ways, and offering opportunities to remove yourself quickly and come back quickly too, to find the space completely different the next time.

Still, the appeal of these types of places is not universal. Some groups of people, for reasons associated with their individualized authenticity as well as for other class-, race- and religion-based reasons, may never find a place that feels like home, here. Nor is placemaking resistant to the demands of capital, which drive up the cost of these places in step with imputed increases to their social value. These are gentrified urban terrains, and from this perspective, the more imageable the neighbourhood, the more tenuous the environment's claim to a unique offering of authenticity. Still, the suite of authenticity offerings in this new urban design has widening appeal in many cities of the world. Indeed, these spaces are so widely appealing that their claim to authenticity is threatened by their very 'imageability' – their predictable public appeal from one city to the next. For urbanites to approach an authentically native identity in this spreading understanding of the ideal urban public space, the material construct of city neighbourhoods, as a set, needs to approach universal imageability and legibility for all. For humanity to become native to the city, and articulate and enact notions of justice and sustainability that fit the urban environment, everyone should have access to understanding its unique worth.

Thus, placemade urbanism raises a number of questions related to its justice outcomes (nor are its environmental sustainability outcomes settled; for more on this, consider: Poumanyvong and Kaneko 2010; Hoornweg et al. 2011; Kennedy et al. 2015). These justice questions need to be resolved. Can a neighbourhood offer an authentic identity, even when the same sort of neighbourhood and the same sorts of habits and freedoms are offered in many other neighbourhoods in many other cities? Can the oppression and exclusion within these neighbourhoods be brought out from the brickwork without threatening the root of the claims they make to the notion of real, native, urban places? Are these neighbourhoods entirely dependent upon neoliberal governance regimes and control by certain forms of capital, and if so, could there be any sense to claims of urban authenticity here? If these sites are designed and intended to permit public discourse, then can we look to this discourse for a view of the potential to expand their inclusionary prospects? Or, in neighbourhoods where identity is forged upon authentic individuality, does the notion of proposing a wider collective sense of community have any bearing at all? We turn to a pragmatic understanding of what happens in a public dispute in order to shed light on some of these questions.

## The dynamics of argumentation

In the city of contemporary celebrations, then, we have a city designed to optimize a sense of authenticity that is nevertheless strictly tied to the demands of neoliberal governance structures and capital accumulation. We have residents for whom this offering of authenticity has strong appeal, but whose sense of authenticity and freedom is compromised by the roll-out of this very imageable urban design, revealing not only the farce of the uniqueness of the current urban living appeal, but also revealing its injustices and exclusions. Finally, we have residents whose customary mode of engaging is based upon a personalized ideal of the planned, self-actualized individual, for whom acting in concert with a community along other lines may present a conflict with their very sense of personal autonomy. These dynamics shape the arguments and resolutions that can transpire. They shape the social interactions among individuals in the formatted environments of the contemporary city.

The key to moving toward engagement, in the pragmatic view, is that one need not abandon one's understanding of justice or authenticity, altogether, in order to reach a pragmatic compromise in a given context. Instead, the pragmatic approach seeks a sense of solidity or continuity, not in individuals as radical self-sufficient beings but in individuals engaging within the physical determinants of context. In the language of *On Justification*, what is crucial to understand in order to advance the prospects for justice in a given dispute is: "the arrangements of objects that qualify the various situations in which persons are acting when they attribute value to these objects" (Boltanski and Thévenot 2006: 216). A self-actualizing, well-planned and liberal-intentioned individual can see each public dispute as demanding a compromise among various, equally worthy individual plans and perspectives on justice, and can be brought to agree to such a compromise without suffering the anxiety of letting go of their own life plan. This kind of reassurance within one's own personality, according to Thévenot (2014: 11), is in fact a prerequisite of effective engagement in our time:

> I conceived the construction of personality – or self – as based on a variety of relations to the world that contribute to self-reassurance... Such self-coordination has to be drawn upon to form expectations about someone else's course of action when coordinating with others.

In the present urban moment, the predictable process of negotiation would appear to make good use of just such a strategy. However, the use to which negotiation tends to be put serves the interest of self-advancement and empowerment as opposed to any more collective idea. Pattaroni (2015: 6), along with Thévenot, characterizes the present liberal order in terms of reliance on a search for standards, detached and abstract principles, interest-based negotiations among divergent opinions held by discrete individuals. To take just one example, the website Transformative Tools (see www.transformativetools.org/) counts more than 60 standard systems of certification of sustainable neighbourhoods, with varying criteria. The nature of a mutually agreeable outcome that could be achieved in this context hangs in between the empowered individuals: "creating commonality then works based on processes of negotiation and the weighing of interests."

It is not difficult to see how this strategy supports participatory and deliberative politics. The City of Vancouver, for example, is not lying when it reports that 35,000 people contributed their views to the creation of the Greenest City Action Plan (Holden and Larsen 2015). It is much more difficult, however, to see how it would support the construction of arguments for rights and justice based upon social ideas that are not currently well represented in people's autonomous life plans, such as the social idea of a "grammar of commonality through affinity" (Pattaroni 2015: 6). When it comes to social ideas of neighbourhood attachment, belonging, responsibility beyond what is enforceable, responsibility to others beyond our personal relations – namely, social ideas upon which just outcomes hang in a democracy – this form of negotiation falls short.

Of course, it is even more complicated than this. The flip side of this situation of multiple arguments and justifications of what is good about the city is that it gives actors the chance to avoid settling their affiliations, to avoid resolving reality tests of any of their propositions, to confound effective argumentation by arguing across multiple modes at the same occasion. This becomes a particularly valuable strategy for those who, recognizing deficiencies in their position in one model of justice, can make claims to worth on the very same basis in a different model. They can, in this way, argue that what is seen as their sloth in another's view, based on their understanding of industry and efficiency as valuable and worthy for attaining justice, can be just as fervently argued as success based upon a view of worth grounded in restraint, sufficiency, and "slow living." Christiania, an autonomous "slow living" neighbourhood in Copenhagen, has taken a strong stance of opposition to the expansion of a cycling super highway through its territory on such grounds (this case is discussed in more detail later on in this chapter). Rather than a context-based compromise, the most that such a dispute may arrive at is manipulative deadlock of differing views of justice.

Let's back up from the pragmatic approach to examine more closely what the importance of the notion of authenticity is to social and political actors in the contemporary city. We do this in order to return, at the close of this chapter, to a pragmatic understanding of the use to which this urban sense of authenticity can be put.

## Our cities become ourselves

Throughout our urban history, writers have given us a range of metaphors by which to understand the city, from machine to be feared, or dominated, to organic or networked system which we might either cultivate differently or rewire, as the case may be. We have seen the city as post-ethical Babylon, as Second Coming Jerusalem, the satanic mills of industrial Edinburgh, the liberatory Garden City, and the futuristic, barrier-free Broadacre City. These days, the surest contemporary urban metaphor that we have is that of the city as ourselves. The new urban quest is to find ourselves by leveraging a home in the city as a kind of city-self authenticity. If the roots of our own true selves do not rouse our passions for the city, what will?

In her 2010 book, *Naked City*, Sharon Zukin asks why urbanites seem to crave "authenticity," a concept she approximates with the characteristics that Jane Jacobs famously valued about cities too, such as density, diversity, character, and liveliness (Zukin 2010). In the new urbanity, Zukin (2009: 544) notices claims to authenticity getting stretched and tied to the "hegemonic global urbanism" that translates into a particular cultural phenomenon of style and experience. No longer associated with origins or roots of individuals or communities, the new authenticity that is the object of urban desire comes without having to be stuck in a single place, or a single identity, for one's entire life. She notes also that "claiming authenticity becomes prevalent at a time when identities are unstable and people are judged by their performance rather than by their history or innate character" (Zukin 2010: xii). That is, within a context of intensifying instability, urban life today offers multiple authentic identities, at once local and global, to those who work at it: a way to be authentically yourself here, and in another way yourself over there, and yet another way yourself by morning. This prospect offers an extremely appealing flexibility in conditions where being footloose with your sense of self seems to offer better survival prospects than does stasis, and more possibilities for self-fulfilment with the next iteration of one's identity, are just around the next corner.

The dark side to this new version of urban authenticity is that it becomes, as Zadie Smith (2014) reflects: "narrow, almost obsessive" – a measure socio-pathic. It is also a modern reflection of a compulsion as old as the liberal, individualist "pursuit of happiness." Smith's Manhattan exemplifies the promise and downfall of urban authenticity; it is: "a perfect place for self-empower-ment – as long as you're pretty empowered to begin with." The narrowing of urban authenticity drawing this objection is a narrowing of urban class diver-sity. If the city offers authenticity to only a select familiar few, what are the rest left to pursue? This is a question of justice. Equally, it is a question of urbanism. In his *Politics*, Aristotle warned that the city should be protected from an over-reliance on familiarity and an under-appreciation of the importance of diversity. Essential to the city was its pluralism: "a city is by nature a plurality, and when its unification goes too far, from a city it becomes a family" (Aristotle 1962 [1885]: II, 2, 85). A city of people who see themselves as belonging to one family is not a city at all, because unification has gone too far. A city of authentic individual narcissists is not a city either, because it is not a plurality.

For Zukin, Jane Jacobs, and others too, authenticity as a concept that goes hand-in-glove with contemporary urban trends offers the potential for far more informal as well as formal social action. In keeping with theorists such as Henri Lefebvre and David Harvey, Zukin sees the right to the city as "a right to change ourselves by changing the city." Because of the density and potential for social mixing, values and demands for all manner of rights clash in the public square. The very basic character of diversity provides a fertile ground for engaging the struggles people wage for human rights and freedoms, in both mundane and more exceptional ways. Zukin (2010: 244) argues that, because authenticity is produced by social relationships, it "must be used to reshape the rights of ownership. Claiming authenticity can suggest a right to the city, a human right, that is cultivated by long-time residence, use, and habit." By this same token, authenticity can wield cultural power to exclude and displace others who do not appear to hold the same rights because of where and how they live, what they look like, and how they spend their time. All the diverse buildings and inclusive public spaces in the urban world will not change this social dynamic, without explicit attention to the need for authentic equity and diversity in law, policy and practice, too. This reflects the views of Iris Marion Young (1990) and Susan Fainstein (2010): for the work of urban authenticity to operate in parallel with the work of justice seeking, institutions need to support a democracy that respects group difference and ensures the participation of marginalized voices in decision-making processes. Such institutions keep a concept of justice based on considerations of equity, diversity, and democracy, at the forefront of decision-making.

This is one crucial step toward justice in sustainable city pursuits. Another is to tackle the compulsion toward standards that compromise between incommensurable views of justice. Is there no insurmountable contradiction in creating places that are at the same time locally authentic and based on a globally uniform design, as imageable in Shanghai as in St. Petersburg? Can we all become interchangeably native and cosmopolitan within the flow of global urban networks and call authenticity an open and democratic concept in so doing? Martin Hajer (1995) refers to this as the aspiration for "zero-friction society." As we will further explain when we expand upon the urban trait of risk resilience in Chapter 6, taking this position glosses over the exclusion of those who do not fit the image, particularly the poor, and it also negates the rooted and personal notion of authenticity.

Today's urbanites demand both authenticities. They travel the city to revisit the memories they have, their connections and impressions, built up over time, and they insist on their right never to be the same way a second time. Our understanding of urban possibilities in relation to this radically new notion of authenticity is based upon an understanding of the city not as a local scale as per Purcell (2006), but more as a 'condition' as per Louis Wirth (1938). The urban idea that motivates our argument here represents more than a scale. To see the urban idea as a new scale of focus, we need only change the lens in our camera, put down the magnifier, and blink our eyes. But if the urban represents not a scale but a condition, there is no tool or cognitive trick to overcome thinking in the urban mode, tying our authentic individuality to our home in the city. The constructive work lies, instead, in better understanding how to put the habits of our urban lives to better, more authentic, more justice-serving pursuits.

## Placemaking in the public domain

Some urban critics cry foul at the very mention of community, claiming that it is a concept entirely bought and sold by capital (Joseph 2002). Others see within the decline of traditional community ideals an emergent hope at the intersection of individualism and social quality. This intersection can be labelled authenticity in community relationships and common spaces. At this intersection, the question that is posed is what it means to be authentically oneself that makes an offering to the spirit of community.

Notwithstanding considerable reasoned doubts about the prospects within the new urban design for the training of social interactions toward such authentic engagements, some have recognized pragmatic potential here too. Powerful among these offerings is Ash Amin's (2008) notion of aesthetic disruption within the public spaces of the contemporary city. Amin calls aesthetic disruption a key offering of the contemporary urban design standards. Such disruptions happen in those active, diversely populated, lively open and public spaces that are crucial to the community-oriented enactment of multiple authenticities. With the public scene ever changing and unfolding, one's authentic belonging and right to act up and act out in public space cannot possibly be based on long-term membership, only on the virtue of being-there-now.

In its ideal manifestation, Amin (ibid.: 16) refers to aesthetic disruption as being put to work "in the spirit of reinventing the ties that bind." This sense of existing or impending aesthetic disruption can be witnessed in the "conviviality" and the "civic remains" of "the direct experience of multiculture" (ibid.: 18–19). Conviviality in the urban mode is experienced as "an important everyday virtue of living with difference ... getting around the mainstream instinct to deny minorities the right to be different or to require sameness or conformity from them" (ibid.: 18–19). It is, moreover,

> experienced as ... a brush with multiplicity that is experienced, even momentarily, as a promise of plenitude ... the gains to be had from access to collective resources, the knowledge that more does not become less through usage, the assurance of belonging to a larger fabric of urban life, perhaps even the knowledge that the space can recover from minor violations.
>
> (ibid.: 19)

(see Figures 4.1, 4.2, 4.3).

For urban public spaces, being open to and actually accommodating an array of users and uses, in a wide range of time, space, and people configurations, is as much a key to their quality as the material infrastructure of the places. Iris Marion Young (1990) offers the notion of the public sphere, a discursive space, essential to democracy, for mediating between strangers where claims and criticisms can be made and heard by many others. Whereas to an advocacy organization like PPS, the spatial dimensions of space are key to its public nature, Young recognized the need for attention to the ideological and political ideas that may not be overtly visible in public space, but may manifest control and oppression just the same. The creation of an open discursive space depends upon work to foreground notions of the public interest, the common good, and community, in contexts of individualism and diversity, asserting the value and possibility of finding common ground in the public spaces of the city.

Doreen Massey (2005) has characterized this kind of open discursive space with respect to the condition of 'throwntogetherness': the opportunity to walk through and mix with a diversity of people and practices, scheduled routines and surprise novelties, with any conscious response by the people so thrown

*Figure 4.1* Mural, Atlantic Avenue, Brooklyn, NY. Visible signs of common place, ordinary justice aspirations for the city, expressed with obvious care and striking talent, can elevate the credibility of a claim to the worthiness of the argument itself.

Source: Photo, the author.

*Figure 4.2* Street poetry on an old mattress, Richmond Street, Toronto. The everyday acts of community, such as disposing of household wastes, in the everyday common spaces reserved for these acts, offer opportunities to advance understandings about a model of justice (here: reuse, thrift, humility, sharing) in ways that also offer an "aesthetic disruption."

Source: Photo, the author.

*Figure 4.3* Aesthetic disruption on 2nd Avenue in Vancouver: Do-It-Yourself Development Permit Application notice board.
Source: Photo, the author, www.OtherSights.ca.

together only ever contingent. Amin (2008: 15) makes a cogent argument about the need for "conditions of plural and inclusive organization of public space." He asserts that public space is made useful in a wide diversity of "urban-life organizing strategies" so as to permit a wide range of situations to arise, and to permit the formation of constituencies for the space. These constituencies, specific to particular public spaces (and, by extension, neighbourhoods), display the "powerful symbolic and sensory code of public culture" (ibid.: 15) and generate the "urban capacity to negotiate complexity" (ibid.: 12). Great public spaces serve these roles while still remaining open to new groups, individuals, and uses, and open in particular to the ongoing prospect of new relationships emerging, surprises, and negotiations. That is, they are non-hierarchical spaces, organized and controlled just enough for people to venture forth into them, but not enough to take away the freedom within for urbanites to enact their right to the collective experience of being human in the city and "a right to change ourselves by changing the city" (Harvey 2008: 23).

Where design-based and political-based approaches to the design of truly public spaces have come together is in concepts like Michael Walzer's (1986) notion of open-minded space, Edward Soja's (1996) notion of thirdspace, and what Groth and Corijn (2005) call "free zones." All of these concepts become particularly operable at the neighbourhood scale. Soja's notion of thirdspace, as distinct from either public space, open to any and all comers, and private space, restricted to particular uses and users, is as a space 'in between' the roles and functions of both. In practice, thirdspaces are neither public thoroughfares, free of any restriction, nor private corporate spaces that extract particular kinds of fees for particular kinds of uses (Soja 1996). They are, instead, places where privacy and familiarity can be achieved, but where few or no restrictions are placed on the kinds of ideas that can be expressed, and no fee for entry or time is charged. Public libraries, certain civic facilities and sidewalk cafés, and some university campus spaces, serve the role of thirdspace.

Walzer (1986) proposes three characteristic kinds of spaces: (1) the intimate space of home; (2) the private space of business and industry; and (3) the public space of citizenship and social congregation. While he distinguishes intimate spaces as a thing apart, he characterizes the private spaces of industry as "single minded spaces," designed and regulated to permit only a single use, usually toward a goal of maximizing efficiency, and public, "open-minded spaces," which are designed and regulated instead to serve a diversity of purposes, according to the wishes of particular users at particular times. Walzer uses "mindedness" as the distinguishing characteristic here, harking back to Young's ideas about how spaces can imply restrictions on ideas and conversations, for the following reasons:

> It's not only that space serves certain purposes known in advance by its user, but also that its design and character stimulate (or repress) certain qualities of attention, interest, forbearance and receptivity. We act differently in different sorts of space, in part, to be sure, because of what we are doing there, but also because of what others are doing, because of what it means to be "there," and because of the look and feel of the space itself.
>
> (ibid.: 321)

In this way, Walzer classifies shopping malls, highways, fast food restaurants, motels and airplanes as single-minded space, whereas city squares or piazzas, city streets, sidewalk cafés and pubs, urban hotels with public rooms connected to the lobby, and trains are open-minded space. While individual tolerances for different levels of 'edginess' and spontaneity in public spaces vary, the fact that widely 'open minded' public spaces do sometimes exist is itself a strong testament to the powerful draw of community life, even in contemporary cities of diversity, flux, and privatization. It is not the naïve perception of power-blind urban designers and place-making advocates that these spaces offer a relaxation of the control of state and capital. It may be, instead, a due recognition of credence to the power of the experience of the collective in public space itself as an additional agent of power, acting alongside power's more often cited modes.

A practical distinction that arises from this classification of private and public spaces is that "open-mindedness requires public subsidy" (ibid.: 328), because of the cost of building and maintaining the quality and safety of these spaces as open and inviting of a wide range of leisure and citizenship activities. Both types of space are necessary for high quality neighbourhood community life. However, the neoliberal critique has clearly identified a decreasing willingness of the public to pay for benefits which do not accrue specifically and individually back to them, squeezing public spaces of their open mindedness (Bresnihan and Byrne 2015). Into this cash-squeezed space, movements to create new forms of reduced public subsidy open spaces have begun in numerous cities. Examples range from the spread of intentional communities, to smaller-scale community gardens in New York and Hong Kong, to new "independent spaces" in Dublin, "social spaces" in Madrid, and "low-cost urbanism" practices (Fournier 2013; Tonkiss 2013; Bresnihan and Byrne 2015). These community-driven efforts operate largely outside of state subsidy, and sometimes entirely outside of the law. They involve varying levels of overt political activity balanced against more pragmatic concerns to find home, workspace, grow food, and opportunities for creative expression and community sentiment. They pool resources communally and work in between the cracks of the capitalist city. While the existence of such counter-examples is exciting, they should not take away from the need for more concerted attention within mainstream urban development budgets and practices to the value and necessity of open minded space.

Without specifying either the markers of the right balance between types of space or the specific stakes involved in failing to achieve that balance, Walzer (1986: 323–324) warns: "the reiteration of single-mindedness at one public site after another seems to me something that civilized societies should avoid." In fact, the mixed-use, urban neighbourhood, with its fine-grained diversity of space, its openness to all people in the public realm, and its manifestation of a potentially wide range of institutional, regulatory, and disruptive social and governance practices, represents a fruitful locus to study the specific quantities and qualities of balance between these three types of space: (1) the intimate, private space of the home; (2) the single-minded space of privately owned places of business; and (3) the open-minded public spaces of the urban ideal.

## Pragmatic opportunity for engagement

Creating and maintaining adequate support for open-minded spaces is not enough to create an engaged and democratic public sphere in which expressing individual authenticities in terms of the public interest establishes a common ground among individuals. For this higher aspiration, we also need to keep justice goals in mind. Thinking back to Hannah Arendt's (1998 [1958]) metaphor of the table as the public realm that both connects those seated at it and prevents them from falling over one another, Mancebo (2015: 24) reminds us that the public realm exists so that people may participate in sustaining it:

> The just city means seating everyone at the table, so that all the inhabitants understand that urban affairs are also their affairs … I don't pretend that seating everyone at the table will suddenly make poverty, segregation and lack of access disappear. It will not. But such an approach – even if insufficient – is the necessary condition to design and carry out a just city.

Growing numbers of cities have expressed policies and practices of public participation and greater inclusion of citizen perspectives in the work of governance. Opportunities for public voice exist in the form of open houses and hearings before city council, for public negotiation and deliberation in the form of citizen assemblies, juries, and participatory budgets, and public involvement in policy implementation through urban design panels and other standing committees, for example (OECD 2009; Galleant and Ciaffi 2014). Some, like the City of Portland (2010), have instituted requirements for staff to document the impact and uptake of citizen engagement initiatives on local government policy and action. Often participation exercises are born of administrative necessity, in local regulation or perception of the need for social license to operate in a community. Much more keenly felt than the push for participation in urban governance contexts, however, is the push for efficiency. In attempting to combine values of efficiency with values of engagement, deliberative processes are typically truncated into much more basic efforts to inject public decision-making with citizen input. Given the desire for participation to result in an influence on decisions, and given the widespread critique of deliberative and partnership-based processes as introducing time inefficiencies that prolong the enacting of decisions, this is not surprising. Overlooked is that the trade-off to efficiency gains in deliberative process design is pragmatic, commons-building, social learning value (see Figure 4.4).

The pragmatic understanding of the purpose of public engagement and empowerment is something different from input to decision-making. It demands a deliberative approach, at arm's length from decision-making. This is necessary for the process to give proper value to the time needed for learning and debate to take place, before any decision is made. This runs the risk of coming to resemble participation-as-talk-shop, distant from any impact on actual governance practices, relegating the social learning that may take place to some loose, indeterminate purpose, seemingly foregoing its force and power (Kanra 2007). Instead, the pragmatic view is that anything less than deliberative process discounts the importance of the diversity that exists in disputes. In committing to begin an engagement, not from an unrealistic assumption of ideal speech conditions, but from given social and material contexts, a process of engagement gives adequate recognition of the real challenges and the real value of diversity and disagreement. A pragmatic approach to engagement, therefore, values social learning over funnelling participants into a single answer, recognizing the overt and covert differences that diversity in identity, in context, and in perspective, make to deliberations. When it comes to accountability, voting may be the democratic gold standard, but other forms of justification also deserve more attention; they make the outcomes of dialogue and learning processes more binding on participants. Pragmatically speaking, the working out in dialogue of one's perspective and why one holds it and being receptive to other's views, may be just as binding in the long run upon those engaged. The hope is that this engagement can lead to compromise among actors able to recognize the difference that their different perspectives make to what is

*Figure 4.4* (a) Participatory planning enhanced with Lego blocks, Vancouver. Urban planners are making attempts to engage citizens in more of a social learning and open-minded mode. An example is the use of arts and crafts as part of a deliberative, participatory exercise. This is a planning event in Vancouver, making use of Lego blocks as well as (b) painting and music as an attractive force to solicit ideas and input into the Greenest City Plan.

Source: Photo, the author.

valuable, and what course of action makes sense across these different under-standings. The rush to a shared decision foregoes the opportunity for learning, shared understanding, and justified compromise in the name of efficiency.

So we have a situation in which even as expectations of participatory politics in the city are now widespread, they have been institutionalized in such a way as to obliterate the public rather than foster the development of urban-authentic publics. Part of the challenge in understanding this disjuncture lies in a poor understanding of the nature and character of "engagement." The institutional frame of participatory politics has become rigid, as new participatory channels are brought to life, with different means of blunting dissent, and "more subtle oppression caused by the transformation of convictions and attachments into deliberative 'opinions' and balanced 'interests'" (Pattaroni 2015: 22). In the same stroke as differences are being brushed over to maintain order, the density of commonality possible between participants is reduced. The trade-off for greater efficiency is the loss of possibility of recognizing the value of place-based, social quality-based attachment. The alternative, what Pattaroni (ibid.: 6) calls the "grammar of commonality through affinity" constitutes a "renewed sharing of the sensitive world" (ibid.: 15) as opposed to an exclusive focus on the interpreted and intelligible world. This way of expressing and creating commonality is essential, as it offers us a "better grasp on the changes in the political forms that occur during urban struggles ... Common places – where experience is pooled – become essential for expressing critiques and inventing new forms of commonality" (ibid.: 6).

Luca Pattaroni offers a pragmatic perspective on this new way of thinking about engagement in the illustrative case of Les Grottes, a neighbourhood adjacent to the Cornavin train station in Geneva. The roots of this neighbourhood lie with a radical and militant social movement that began in the 1970s as an occupation, an assertion of the right of poor residents to stay in an area where urban redevelopment had been prohibited, accompanied by a radical critique of capitalism. The voice of the squatter community grew louder as economic conditions changed for the city and new development was proposed. While much of this redevelopment did push ahead, some of the squatted settlement was retained and in 2009, the neighbourhood residents signed a formal Neighbourhood Contract with the City of Geneva. This signing represents, on the one hand, an abandonment of the long-standing battle of poor residents against the power of the state for the right to remain. On the other hand, though, the Neighbourhood Contract also offers the remaining residents a right to remain in their quarters and a means to guarantee deliberative, participative process in the development of public spaces in the neighbourhood. The older theatre of protest has been lost, for the moment, but a new form of politics of difference has been won at the same time.

Pattaroni proves this by demonstrating how the existence of the Neighbourhood Contract does not quash forms of protest and resistance that are parallel to and completely separate from the contract, signalling the continued existence of the neighbourhood whose residents are able to revisit and keep alive their local political history of radicalism, and it does not silence residents who want to talk about matters that reach well beyond the development, design, and management of the public space that the contract was designed to address. Instead, it represents a new form of compromise in which "people and communities navigated and found a balance between these two extremes" (of militant radicalism, on the one hand, and gentrification, on the other) (ibid.: 11). Rather than a case of the death of radicalism at the hands of capital, the case of Les Grottes reveals a new political arrangement of critique, emerging to fit the particular diversity and arrangement of the moment. As an open-minded space, the purpose of a political compromise is not to see what meagre threads of half-hearted critique may be available in a world in which a neoliberal understanding of worth and value is a foregone conclusion. It is, instead, a creative, empowering experiment. The tool that was created, via the Neighbourhood Contract, has the potential to effectively expose contradictions in the institutionalized understanding of worth and value, such as the institutionalized understanding of the market value of a neighbourhood, and to offer new understandings and solutions from this new, common place. Key to this new practice of resistance and critique is recognition of local, neighbourhood-specific diversity, authenticity, and the "valorization of attachments" (ibid.: 18) beyond the realm of market value. The case of Les Grottes is one of the creation of a new form of struggle, based upon a "multitude of material, conventional and institutional measures used to organize the city" into "a new common place" (ibid.: 4).

A similar case is the Christiania autonomous zone in Copenhagen, also established in the 1970s as an occupation with a radically anti-market view of community development and community life. While this arrangement was tolerable from the point of view of government for the first generation, more recently the government has issued an ultimatum to the district residents: buy the land outright, or stand by for the government to develop it according to the same rules applied elsewhere in the city. Christiania took neither option. Instead, following a multiple-year deliberative process within the neighbourhood, not to mention a struggle within the court system, the residents arrived a compromise called *folkeaktie*, or People's Share. The campaign amounts to a crowd-sourced fundraising effort, offering investors to be part of protecting this district and its unique values from speculation and marketization, to have a share of "keeping Freetown free" (Christiania Shares 2014). The identification of the values and understandings of worth in Christiania is not dissociable from the space the district occupies: freedom of expression in an open, consensus-based neighbourhood-scale democratic system, freedom of artistic expression above market-based ability to pay, and freedom to live a "slow" life within the city (Winter 2016). Within the contemporary urban political moment, the city's ultimatum to Christiania and Christiania's creative response represent an opportunity to compromise, to create and communicate and enact a new relationship, that might not have existed in a simple maintenance of the status quo.

From a pragmatic view, the guiding rule for engagement is neither poetic expression nor material production nor revolution, but rather "the creative solution of problems by an experimenting intelligence" (Joas 1993: 247–248). The experimenting intelligence takes the shape of a collective, rather than a chain or sequence of individuals. The work is done by "human beings who collectively recognize and discuss their earthly problems and creatively solve them" (ibid.: 257). We advance this intelligence to the degree that we relinquish faith in the force of strategy alone within social processes, and overcome our sense of despair with respect to moving through conflict productively, recognizing instead an indeterminate notion of sociality with a human drive for creative accomplishments. It is a skill to be acquired, and a time-consuming one at that, which also requires a context in the urban environment and the social world that inspires the creation and recreation of worlds that have meaning to those who inhabit them. We need more opportunities to practice this kind of engagement.

This notion of engagement takes a turn toward John Dewey's argument for radical democracy, toward a charge addressed by Michael Walzer (1983) in his thesis of complex equality and toward Iris Marion Young in her thesis of intersecting voices. Young (1997), for example, wrestled with the pitfalls of deliberative democracy (contra Jürgen Habermas and John Rawls) for its narrow focus on rational decision-making (and on ideal theory). As an alternative, she proposed a communicative democracy model responsive to a heterogeneous polity. In the language of contemporary diversity theory: political actors become "culturally competent" in different modes of speaking and listening, what Young calls *intersecting voices*. More than any other contextual factor, the condition of intersecting voices is likely to occur in the city, making this again essential to an urban model of justice. A synthesis of these perspectives on engagement may offer significant value beyond the limits of the original American pragmatic emphasis on communities of inquirers that are willing and able to model themselves upon scientific communities, and beyond the limits of the original works within the pragmatic sociology of critique which limit themselves to public, rational dialogue processes. This understanding of different forms of engagement for different contexts may be well suited to seizing the power of urban, political moments for change.

The local trap does occasionally rear its head. A place-based attachment model of politics has its own oppressive dimensions, and the question of how an experience rooted in a given place becomes a broader political issue is unresolved (Breviglieri and Trom 2003; Thévenot 2006; Cefaï and Terzi 2012). The proliferation of difference in expressions of interests and attachments is perhaps the opposite extreme of the problem of fitting all forms of protest under the rubric of Marxism in the 1960s and the 1970s. We are attached to things and modes that are not place-specific, too, and these different attachments pull us away from neighbourhood-scale politics. Proximity is only one of many motivating factors toward political engagement. But it is an essential motivating factor.

The shift in discourse represented by the pro-urban political idea is empowering and motivating in this context, so long as notions of 'innovation' remain scaled as they are now, ultra locally. So long as it remains widely understood that crucial innovations happen not in board rooms, international convention centres, and city hall chambers, but also in basements, garages, coffee shops, and street corners, people can persist in their belief that what they may achieve working in their own neighbourhoods and cities may matter, may make a difference. Many people continue to be willing to commit to the local scale, and numerous neighbourhoods may be small enough to resist the corrupting influence of scalar politics.

## Conclusion

In this chapter, we have considered the imperative of engagement in urban public space for a move toward sustainability and justice, from a starting point of individual authenticity. The urban design and planning professions have found a strong resonance with this authenticity imperative, inventing an environment for authentic urban living that has evolved into a dominant, desirable global model for urban neighbourhoods today. Urbanites have responded with their feet, flocking to these neighbourhoods where they can afford it. There may well be positive implications for urban sustainability in these designs, at least in contrast with modernist cities and suburban sprawl. The implications for justice, however, are ominous, as pointed out by the local trap presented in Chapter 3. In pragmatic terms, the regime of familiar engagement, that is best suited to the demonstration of authenticity, is counter-productive when it comes to building common ground in urban community. To the cognitive dissonance thus produced by individualists engaging within 'placemade' spaces, I offer the potential return to consideration of the public interest via the concept of open-minded space. The openness of placemade urban neighbourhoods to diverse mixing of uses, and the openness of urbanites to diverse assertions of equally authentic personalities, should create a receptiveness as well to political openness, and the need for aesthetic disruptions as part and parcel of the continuing reinvention of the authenticity of these spaces. Authenticity may, in this way, carry a pragmatic political load. It remains hampered, however, by the problematic position of the notions and experience of community and empowerment in our cities. It is to these justice-defying tensions that we turn next.

## References

Amin, Ash. "Collective Culture and Urban Public Space." *City* 12(1) (2008): 5–24.

Arendt, Hannah. *The Human Condition*. Chicago: University of Chicago Press, 1998 [1958].

Aristotle. *Politics*. Trans. Benjamin Jowett. London: The Clarendon Press, 1962 [1885], pp. 258–259.

Beauregard, Robert and Anna Bounds. "Urban Citizenship." In *Democracy, Citizenship and the Global City*, edited by Engin Isin. New York: Routledge, 2000, pp. 243–256.

Boltanski, Luc and Laurent Thévenot. *On Justification: Economies of Worth*. Princeton, NJ: Princeton University Press, 2006 [1991].

Bresnihan, Patrick and Michael Byrne. "Escape into the City: Everyday Practices of Commoning in the Production of Urban Space in Dublin." *Antipode* 47(1) (2015): 36–54.

Breviglieri, Marc and Danny Trom. "Troubles et tensions en milieu urbain: les épreuves citadines et habitants de la ville." In *Publics politiques, publics médiatiques*, edited by Daniel Cefaï and Dominique Pasquier. Paris: PUF, 2003, pp. 399–416.

Cefaï, Daniel and Cedric Terzi (eds.) *L'expérience des problèmes publics: Perspectives pragmatistes* no. 22. Paris: Editions de l'EHESS, 2012.

Christiania Shares. "Christiania Folkeaktie." Available at: www.christianiafolkeaktie.dk/ 2014 (accessed 18 August 2016).

City of Portland. *Public Involvement Principles*. Portland, OR: City of Portland, 2010.

Fainstein, Susan. *The Just City*. Ithaca, NY: Cornell University Press, 2010.

Fournier, Valerie. "Commoning: On the Social Organisation of the Commons." *Management* 16(4) (2013): 433–453.

Galleant, Nick and Daniela Ciaffi. *Community Action and Planning: Contexts, Drivers and Outcomes*. Chicago: University of Chicago Press, 2014.

Geddes, Patrick. "Civics as Applied Sociology." In *The Ideal City*, edited by Helen Meller. Leicester: Leicester University Press, 1970 [1905].

Groth, John and Eric Corijn. "Reclaiming Urbanity: Indeterminate Spaces, Informal Actors and Urban Agenda Setting." *Urban Studies* 42(3) (2005): 503–526.

Hajer, Martin. *The Politics of Environmental Discourse: Ecological Modernization and the Policy Process*. New York: Oxford University Press, 1995.

Harvey, David. "The Right to the City." *New Left Review* 53(8) (2008): 23–40.

Holden, Meg and Majken Larsen. "Institutionalizing a Policy by any Other Name: In the City of Vancouver's Greenest City Action Plan, Does Climate Change Policy or Sustainability Policy Smell as Sweet?" *Urban Research & Practice* 8(3) (2015): 354–370.

Hoornweg, Daniel, Lorraine Sugar and Claudia L. T. Gomez. "Cities and Greenhouse Gas Emissions: Moving Forward." *Environment and Urbanization* 23 (2011): 207–227.

Jackson, Wes. *Becoming Native to This Place*. Lexington, KY: University Press of Kentucky, 1994.

Jacobs, Allan B. *The Good City: Reflections and Imaginations*. New York: Routledge, 2011.

Jacobs, Jane. *The Death and Life of Great American Cities*. New York: Random House, 1961.

Joas, Hans. *Pragmatism and Social Theory*. Chicago: University of Chicago Press, 1993.

Joseph, Miranda. *Against the Romance of Community*. Minneapolis, MN: University of Minnesota Press, 2002.

Kanra, Basra. *Binary Deliberation: The Role of Social Learning and the Theory and Practice of Deliberative Democracy*. Helsinki: European Consortium for Political Research, 2007.

Kennedy, Christopher. *The Evolution of Great World Cities*. Toronto: University of Toronto Press, 2011.

Kennedy, Christopher, Iain Stewart, Angelo Facchini, *et al.* "Energy and Material Flows of Megacities." *Proceedings of the National Academy of Sciences* 112(19) (2015): 5985–5990.

Lippmann, Walter. *Public Opinion*. New York: Harcourt, Brace, 1922.

Lynch, Kevin. *The Image of the City*. Cambridge, MA: MIT Press, 1960.

Magnusson, Warren. *The Politics of Urbanism: Seeing Like a City*. New York: Routledge, 2013.

Mancebo, François. "Ceci n'est pas une pipe: Unpacking Injustice in Paris." In *The Just City Essays: 26 Visions for Urban Equity, Inclusion and Opportunity*. Edited by Toni Griffin, Ariella Cohen, and David Maddox. V.1. New York: J. Max Bond Center on Design, Spitzer School of Architecture, City College of New York, Next City and The Nature of Cities, 2015, pp. 21–25.

Massey, Doreen. *For Space*. London: Sage, 2005.

OECD (Organization for Economic Cooperation and Development). *Focus on Citizens: Public Engagement for Better Policy*. Paris: OECD, 2009.

Pattaroni, Luca. "Difference and the Common of the City." In *The Making of the Common in Social Relations*, edited by Alexandre Martins and José M. Resende. Newcastle: Cambridge Scholars Publishing, 2015.

Poumanyvong, Phetkeo and Shinji Kaneko. "Does Urbanization Lead to Less Energy Use and Lower $CO_2$ Emissions? A Cross-Country Analysis." *Ecological Economics* 7(2) (2010): 519–529.

Purcell, Mark. "Urban Democracy and the Local Trap." *Urban Studies* 43(11) (2006): 1921–1941.

Smith, Zadie. "Find Your Beach." *The New York Review of Books*, 2014 (Oct. 23). Available at: www.nybooks.com/articles/archives/2014/oct/23/find-your-beach/ (accessed 18 August 2016).

Soja, Edward. *Thirdspace: Journeys to Los Angeles and Other Real-and-Imagined Places*. New York: Wiley-Blackwell, 1996.

Thévenot, Laurent. "Which Road to Follow? The Moral Complexity of an 'Equipped' Humanity." In *Complexities: Social Studies of Knowledge Practices*, edited by John Law and Annemarie Mol. Durham, NC: Duke University Press, 2002, pp. 53–87.

Thévenot, Laurent. *L'action au pluriel: sociologie des régimes de l'engagement*. Paris: La Découverte, 2006.

Thévenot, Laurent. "Voicing Concern and Difference: From Public Space to Common Places." *European Journal of Cultural and Political Sociology* 1(1) (2014): 7–34.

Tonkiss, Fran. "Austerity Urbanism and the Makeshift City." *City* 17(3) (2013): 312–324.

Walzer, Michael. *Spheres of Justice: A Defense of Pluralism and Equality*. New York: Basic Books, 1983.

Walzer, Michael. "Pleasures and Costs of Urbanity." In *Metropolis: Center and Symbol of Our Times*, edited by Philip Kasinitz. New York: New York University Press, 1986, pp. 320–330.

Winter, Amanda K. *Unmasking the Green City: Contested Sustainability in Copenhagen and Vancouver*. Budapest: Central European University, 2016.

Wirth, Louis. "Urbanism as a Way of Life." *American Journal of Sociology* 44(1) (1938): 1–24.

Young, Iris M. *Justice and the Politics of Difference*. Princeton, NJ: Princeton University Press, 1990.

Young, Iris M. *Intersecting Voices: Dilemmas of Gender, Political Philosophy, and Policy*. Princeton, NJ: Princeton University Press, 1997.

Zukin, Sharon. "Changing Landscapes of Power: Opulence and the Urge for Authenticity." *International Journal of Urban and Regional Research* 33(2) (2009): 543–553.

Zukin, Sharon. *Naked City: The Death and Life of Authentic Urban Places*. New York: Oxford University Press, 2010.

# 5 Empowerment in urban communities, after abandoning Utopia

The city that so many join together in celebrating is a place in which liberal, self-seeking individuals imagine themselves free to pursue life plans and to represent themselves as authentic. I am one of them. I grew up listening to "Free To Be You and Me," a popular children's album by Marlo Thomas (1972). Growing up, I was taught that to treat other people with respect was to treat them as if everyone was a unique, ultimately inscrutable, individual. The expectation that follows is that acting as an individual is worthy of respect and that respect adheres most strongly to those who fulfil their promise as individuals. This expectation is so basic that it would be unthinkable that acting out one's individual life plan might at some level be inconsistent with improving the city's sustainability or justice prospects. We should be able to walk or bicycle to work and school, exchange sociable words with our neighbours when watering the community garden or taking a child to soccer practice at the neighbourhood park. We can live well, materially, with a smaller house than our parents or colleagues who live in suburban environments, and a smaller daily travel footprint than those who live outside the city, too. We can make eye contact, at street level, with all manner of people with whom we share the street, and give some of them a dollar, some of them a smile, one an extra pair of sunglasses we no longer had use for. The choice is up to each of us, at each new juncture.

As per Kevin Lynch (1960), there is a new "imageability" to the city for those coming home to such environments. It is a welcoming imageability, with a human scale. It is not overly demanding of us in terms of consistency to any particular vision of selfhood or citizenship. It also has blind spots in terms of sustainability and justice. At least, from a pragmatic perspective, it offers a situation in which residents may be prepared to consider compromises with others in the service of a better outcome, because doing so does not sacrifice an essential component of their authenticity. As we discussed in Chapter 4, this personal stance of willingness and ability to consider compromises is key to the opportunity hidden within these placemade urban environments to better serve the public interest. Or, at least, to better align it with an aspiration to sustainability and justice than is the case today.

Lewis Mumford devoted his 1922 book, *The Story of Utopias,* to characterizing varieties of utopian ideals. He investigated why utopian visions are both necessary and ridiculous – necessary because they make the world tolerable, ridiculous because of the odds stacked against them ever seeing the light of practice. Mumford considered this human search for utopia to be more likely in the city than in any other kind of place. Cities offer capacity in a world of disempowerment and a vision of authentic life that can be pursued even through extreme disillusion. In *The Prospect of Cities*, John Friedmann (2002) grappled with the question of the good city and how far we can move toward attaining such a city by envisioning it in an ideal form. In his view, generating and sustaining multiple authenticities are key to achieving a balance between the vision and the practice of utopianism in cities. He takes the critical pursuit of utopia as a necessary task of urbanists, in the same stream as Mumford, who aimed to unite the ecstasy of good vision with the science of practical advancement. Utopian ideas are necessary because they give us our only available alternative to nothingness. They are ridiculous to the extent that they stop short of a full union of idealism and science, meaning a marriage of the vision of the ideal to be achieved with the sustained energy for change and collective skills to bring to implementation.

What I propose in this chapter, by examining the notion of empowerment and its application to the service of community-building in this new urban order, is that the contemporary urban moment makes a different kind of utopianism possible. A pragmatic, anti-utopian utopianism that is in tune with the demands of contemporary urban life and engagement as it seeks inspiration and elevation of ideals.

The imageable new urban utopia I have in mind has already been described in this book as the new urban model of 'placemade' neighbourhoods. These do not find their expression in a Garden City, ideal human settlement planned to the detail of private residence to greenspace ratio and positioning, population, occupation mix, mode of governance, and much more (Fishman 1982). Instead, they take shape in the bricolage designs of Jan Gehl, the Project for Public Spaces, Jane Jacobs' Greenwich Village, and the work of other leading placemakers. They have some semi-autonomous combination of neighbour-hood-scaled smart growth and transit-oriented development strategies, medium-density, attached homes that provide an eye on the street and a patch of private ground from which neighbourhood interaction can start, green building standards, green markets and green spaces, commuter cycling challenges, car sharing and all manner of mobile technology-facilitated new cooperatives. It would be a mistake to read this variety of urban utopias as incrementalist. It manifests, rather, as a sort of do-it-yourself social transformation in tiny individualistic bites by authentic people and planners, with their own hedonistic motivations. The enthusiastic adoption of these principles by cities and individuals alike, without the presence of a master architect, is what differentiates the contemporary urban utopias from older utopianisms.

To be generous, these numerous innovations in urban middle-class living address sustainability and justice concerns in a typical, happenstance, piecemeal fashion. One summer, there may be a zero waste competition. Then, an artisan baker who employs street-involved youth on a trial basis. A public art committee might pop up with a piano to donate, and a grant from the paint store to donate to the high school for a mural project. Many of these experiments may fail, or fail to live up to a commitment to justice and sustainability over time. But in the critically charged environment of cities, even our urban experimental failures can add to the texture and conflict that incline us toward understanding a better, uniquely urban, future that is just and sustainable. In order to understand how this could be the case, we need to grasp the pragmatic perspective on empowerment and community. Within the utopian city put forward by the contemporary urban "celebration" is an implicit "celebration" of local empowerment.

Public participation and empowerment have indeed formed a more considered part of urban and neighbourhood development planning and programming in recent years, and are often reasoned to be of central importance, but with ambiguous and often disappointing results in practice. Generating effective, empowered engagement of the urban community is more than a question of creating opportunities for input, in the pragmatic view. This outcome, instead, demands an investment in long-term social learning to regenerate a cultural interest and capacity in the common good. To pursue this path is to trace the course of transition from the authentic individual toward the Great Community. The Great Community ideal, envisaged by pragmatist John Dewey (1981 [1927]), is the supportive structure of democracy that sits behind a democratic public, a sentiment of the common good that presupposes a city of spaces that people hold together through their talk and actions. This is the basis for a pragmatic belief in the possibility, at this time, for a re-appropriation of the notion of citizen empowerment, in terms of a greater "commoning" of the city, a move toward justice and sustainability.

## From the Great Society to the Great Community

Arguments about the unity and diversity in the common good of the utopian city have risen and fallen since the advent of modern society. The classic liberal democratic formulation of the societal ideal is that of the Great Society. Crafted by US President Lyndon B. Johnson into a package of domestic reforms whose goal was to end poverty and racial discrimination, the notion of the Great Society has long inspired the thought and passion of public intellectuals. The notion saw the structural unit of the nation state united with democratic values, and a common evolution of both structural and value-based forms of society.

However, as early as the 1920s, the Great Society was recognized to be an unattainable ideal. Acknowledging that contemporary conditions resembled "a most inadequate picture of the Great Society," Walter Lippmann (1922: 36) was pessimistic about the prospects of the concept to come to greater fruition:

> It is no wonder that moral judgment is so much more common than constructive thought. Yet in truly effective thinking the prime necessity is to liquidate judgments, regain an innocent eye, disentangle feelings, be

curious and open hearted. Man's history being what it is, political opinion on the subject of the Great Society requires an amount of selfless equanimity rarely attainable by any one for any length of time. We are concerned in public affairs, but immersed in our private ones. The time and attention are limited that we can spare for the labor of not taking opinions for granted, and we are subject to constant interruption.

In more contemporary times, agonists and postmodern relativists have criticized the Great Society ideal, based on recognition of the value of difference between members of a democratic public, making "free and full intercommunication" both undesirable and oppressive. From another angle, critics of neoliberalism have identified the multiple and unstoppable means by which the drive for capital accumulation affects all societal interactions, making an interaction in which a Great Society was either the goal or the actual outcome inconceivable. From a Foucauldian governmentality perspective, the rhetorical move of promoting a Great Society could be taken as a move toward a soft approach to imposing state power at the scale of the community without this exercise of power being perceived as overtly threatening freedom. Thus, actors favouring state power would naturally seek to promote slippage between structural and social approaches to understanding community because such slippage would reinforce the role and extend the reach of the state. Nikolas Rose (1996: 331) notes how "the social" may give way to "the community" as a new territory for the administration of individual and collective existence:

> [I]t seems as if we are seeing the emergence of a range of rationalities and techniques that seek to govern without governing *society*, to govern through regulated choices made by discrete and autonomous actors in the context of their particular commitments to families and communities.

Walzer (1986) illustrates this shift by pointing out the replacement of a governance commitment to *welfare*, which has particular financial and rights-based implications, with a commitment to *wellbeing*, which lacks such specific responsibilities. This shift can be seen as reflecting and reinforcing the success of liberal individualism, which obliterates older aspirations for the Great Society with the promise of individualistic gratification. Whereas welfare demands too much from liberal individuals in terms of belief in the possibility of a large social group, acting in concert and mutual benefit, wellbeing has a more familiar, individualist ring to it. A right to wellbeing supports might be extended to an expanding community if this is the wish of the individuals within that community, but inclusiveness is not key to meeting a wellbeing ideal. Wellbeing-seeking individuals can realize their life goals without risking it all for the ability of others to realize theirs:

> Increasingly, we conceive of well-being exclusively in terms of the self. Liberalism breeds an expansive desire for comfort and closeness, useful

commodities and loving persons, while an older republicanism, historically associated with open minded space, provides us only with monuments and fellow citizens.

(ibid.: 325)

In his own critique of the Great Society, Dewey specified the failing of the Great Society as insufficiently attentive to the demands of democracy at the community scale. The slippage of the notion of the common good toward the individualized, competitive good and of the notion of community away from democratic participation and toward a streamlined implementation of governance is part and parcel of this insufficiency. Dewey drew a key connection between the Great Society and what he called the Great Community. Dewey's Great Community could only occur if citizens, experts, and decision-makers had "free and full intercommunication" (Dewey 1927: 211), and an active public life in which to maintain this communication: "The clear consciousness of a communal life, in all its implications, constitutes the idea of democracy" (ibid.: 149). To Dewey, democracy was the most fulsome and valuable ambition of society and represented a marriage of communitarian and liberal-individualistic ideals, along with scientific justifications of the highest and best purpose of society. Moreover, Dewey's ideal of democracy was a participatory ideal, which both created and depended upon a fully informed public capable of dialogue on all socially important decisions. As such, it was meaningless to speak in a social and political sense of a democratic public in the absence of a precondition of community wellbeing. To Dewey (ibid.: 142): "Till the Great Society is converted in to a Great Community, the Public will remain in eclipse."

The need to justify arguments made and actions taken in the public sphere, using argumentation that is acceptable as legitimate by a democratic populace, is key to the entire pragmatic framework of reasoning about justice. Equally, the value of understanding the role of argument in the production of policy learning toward urban sustainability has been noted by Bulkeley (2006). At the most basic level, justifications offer respect in communication and invite scrutiny and accountability. As Beauregard (2015: 48) puts it, "The statement 'we did this because' shows respect for others." The next step in offering a pragmatic justification is ensuring that this justification is based on an understanding of the generalized common good. Such a reason is essential because private or narrow group interests could never be accepted by a democratic populace as a whole as legitimate (see Figures 5.1 (a) and (b)). The public nature of the justification is crucial in order to ensure that the argument is also open to critique and counter-claims from any interested member of the democratic populace. Publicness also ensures that the justification comes with proof that is in keeping with the model of justice, and that this proof is also available for public scrutiny, testing, and offering of counter evidence. The public nature of disputes included in this framework, plus their necessary reference (implicit or explicit) to a notion of the common good, opens up the possibility that disputes will result in action by collectivities to counter domination. The possibility also exists, within this way of understanding public argument and action, to grasp an evolving sense of the constituents of and adherents to public legitimacy. This is a key link to the American pragmatism of John Dewey, as evidenced most directly in his notion of the community of inquirers. The practice of this form of analysis also follows a pragmatic notion of truth, in which truth must be demonstrated not by recourse to first principles but "by pointing to the real world" (Thévenot et al. 2000: 238).

Distinct from a Habermasian communicative action approach (Habermas 1984), the pragmatic approach does not impose a single ideal set of conditions for communicative rationality, but attends to the reality of diverse social and political settings and the material conditions on which actors rely when they form their opinions and arguments. Moreover, a pragmatic approach recognizes that actors engage a variety of capacities in the public sphere and may act in a different capacity each time depending on what community membership or perspective takes precedence for them. Rather than think of a communicative action ideal, this approach seeks to "acknowledge actors' realisms, meaning the plurality of ways they engage with an appropriate reality to put their capacity into action" (Thévenot, in Blokker and Brighenti 2011: 12). In essence, there is more to understanding public debates, disagreements, and the engaged work of community building than moving from the specific to the general, from the individual to the collective, or from the emotional to the scientific (Boltanski and Chiapello 2005 [1999]; Boltanski and Thévenot 2006).

Communities place boundaries on the kinds of justifications that can be offered legitimately in various situations and conditions. For example, where a debate focuses on the best means of housing and caring for the poor, a 'good' justification in most cases cannot merely appeal to blatant private interests of the well-off, but must respond to an established general interest in common social value. Of course, a range of legitimate justifications can still exist, and these will continue to be disputed in democratic society.

Take, for example, the two contrasting cases of housing the hard-to-house in False Creek South, a neighbourhood constructed in Vancouver in the 1970s, and in Stockholm Royal Seaport, a major contemporary redevelopment project (Figures 5.2(a) and (b)). The False Creek South neighbourhood of Vancouver is a positive case in point of the translation of a social value argument about pluralism, diversity, and inclusion into neighbourhood design. Built in the 1970s on public land, just around the peninsula from Vancouver's downtown, False Creek South is Vancouver's original family-oriented, medium-density development. The design consists of mid-rise apartment buildings and townhouses, with front doors oriented toward the seawall and its active transportation routes, including routes to the neighbourhood school and the Granville Island public market. The primary tenure models are unit ownership on leasehold land (about a third), subsidized social housing units (about a third), and cooperative family housing (about a third). This design was strongly justified in the plan issued by the False Creek Development Group plan. It reasoned:

*Figure 5.1* Who are they kidding? (a) Future home of a community park, Vancouver; (b) local business happening here, Victoria, British Columbia. Attempts to introduce new models of justice, such as a roadside sign boasting of authentically local development, that do not fit their material context, such as a gated field, are not likely to change the course of solutions-seeking in a city. The argument about the value being pursued – local amenity and economic development – is not recognizably worthy in this local environmental context.

Source: Photos, Michael Wakely and Meg Holden.

If the urban dweller's mental and emotional needs are to be met, then all parts of the social community must have as wide a range of diversity as possible. Diversity in lifestyle, household type and income level should be able to be experienced at all scales within the urban environment and by all members of the community. You should be able to experience diversity when you walk around. Your children should be able to experience it as they are growing up.

(False Creek Development Group 1974: 8)

The justification is clear, and should be striking to the ears of contemporary urbanists: Communities which offer little social and physical diversity are unhealthy. People living in them have limited freedoms and limited access to the essential experiences of urban living, namely, the experience of a wide range of enacted values, habits, and beliefs. The children of middle-class people suffer if they do not experience diversity in lifestyle choice, household type, and income, in their home environments.

This case stands in contrast to the arguments made more frequently in new urban development contexts today, which tend to focus on the most efficient use of limited public land and resources. Take, for example, the justification offered in the major Royal Seaport development in Stockholm against integrating social housing into this, one of the largest urban redevelopment projects in Northern Europe. In the 1960s and the 1970s, while the Canadian government was investing in housing projects like False Creek South, Sweden was building "the Million Programs": massive prefabricated housing projects on the urban outskirts, to service the social housing needs of (aspirationally) a million residents. In the Stockholm Royal Seaport (SRS), currently under development, no new housing is targeted for this same demographic of housing need. The justification offered by staff from the public development authority, who gave me a tour, comes from an entirely different direction than that advanced in False Creek South. A small piece of our conversation tells the story:

MEG: But if the City is the landowner, and if the City can set the price at which it sells the land and make the housing that results much less expensive, why can't the city do something about improving affordability here?

SRS: Because if they did this, then there would not be enough money to do important works for everyone like building new metro lines and roads.

MEG: So not having social housing here is a political position?

SRS: Yes, in a way. If someone is homeless, they can go to the social service worker, and they will find him a home. It just won't be here.

## Reviving public participation to serve sustainability and justice

A radical, democratic model of engagement is frequently articulated as part of the new urban utopia. It is a vision that goes something like this:

*Figure 5.2* (a) Streetscape in Vancouver False Creek South; (b) Stockholm Royal Seaport
Source: Photos, the author.

The just city of the new century will be a city in which decision-making
processes are not monopolized by few "representatives" and political parties, but are in the hands of the communities and the citizens; the land, the
infrastructure, the facilities and the public and private resources are

distributed for social use and enjoyment; the city is recognized as a result of the productive contributions of the different actors; and the goal of the economic activities is collective wellbeing; all human rights are respected, protected and guaranteed for everyone; and we conceive of ourselves as part of nature, and nature as something sacred that we all should take care of.

(Zarate 2015: 29)

Urban designers and planners behind the placemaking movement assert that the city offers nothing but support of this vision. To be fair, a thread of the urban design and planning literature does include a focus on the role of local democratic practices for wellbeing. Many advocate for public participation processes as a means to give both short-term importance to the process and concrete outcomes of the matter in the process itself, and long-term importance to offering experiences, skills, and empowerment that participants can carry forward with them into further collective action (Eicher and Kawachi 2011). Well-designed and well-equipped public spaces are also and always socially and politically open to an array of users and uses, in a wide range of time, space, and people configurations. Their openness is a key design quality, and a challenge is put to their governance to maintain this.

Effective public participation has also been articulated as a goal of sustainable development since the 1992 Rio Declaration, Principle 10 of which refers to "the participation of all concerned citizens, at the relevant level," coupled with "appropriate access to information," and efforts taken by the state in favour of "public awareness" and "effective access to judicial and administrative proceedings" (United Nations 1992). A variety of benefits are typically associated with public participation for sustainable development, namely: an increased willingness to change, increasingly informed public, reduced disaffection and defused anger, stronger emphasis of policies on local benefits, more efficient use of government resources, particularly with respect to implementation, an injection of creative ideas, and improved communication and decision-making (Bell and Morse 2001; United Nations 2002; Cowell and Owens 2006; Raco 2007). Successful models of local democracy and community engagement in planning are now thought to foster social capital, promote local ownership and engender accountability (World Health Organization and United Nations Human Settlements Programme 2010; Aboelata et al. 2011). Others have raised questions about the likelihood of any of these results (Mouffe 2000; Fung and Wright 2003; Holden 2011).

While community participation in urban governance can empower communities and address injustices, poorly managed *ad hoc* participation processes can contradict these aims (Katz et al. 2015). In Chapter 4, we discussed the problem of rushing to decision-making as a point of failure for a participation process. Quarter and Mook (2010) offer a range of frequent flaws in attempts to engage and empower the public in such processes, including: the treatment of citizens as clients or customers, inauthentic citizen engagement efforts that do little to establish channels for making use of the engagement offered, and a lack of disclosure and transparency. Nor can the quality of political engagement be reduced to questions of input opportunities alone – effective work also needs to address the urban civic culture, the human political and social community, and the practice of citizenship as a mode of engagement. At the same time, current work has revealed the challenges of "professional citizens" dominating some open public participation processes (Michels and de Graaf 2010).

Whether considered as a design challenge or a governance challenge, generating genuinely engaged citizenship is more work than it might seem. Recognition of open engagement of citizens in the design and management of communities as a civic right is an even greater challenge. "Good" physical design can even be perceived as part of the problem. Neighbourhoods with a surplus of attention to their design can project a sense of completeness that detracts from the perception of a need for residents and occupants to engage. Without sustained engagement of empowered urban citizens, the potential to capture the opportunity for the just city is lost. With it is lost the potential of open discourse to generate new ideas about better uses of city spaces, infrastructures, services and communities, toward more ambitious sustainability and justice goals. The work of transformation needs to be pursued within social relationships and habits of interaction just as much as it needs to be pursued in environmental relationships and relationships to capital. What's more, the connections need to be made via experience and reflection.

## Learning socially to engage in communities

The question of bringing the public back into the daylight, out of its eclipse, in order to articulate a common ground between the community and the individual ideal, is a question of social learning. Kevin Lynch (1981: 97) referred to cities as the products of thinking human beings who are capable of learning; others have called the city a "learning ecology" (Yanarella 1999: 211). Social learning can be thought of as any process that takes place in a social context, that is, engaging more than one individual, in which critical questions are raised, and attempts are made to answer them (Dryzek and Niemeyer 2010). This concept of an urban scene pregnant with learning potential provides the very basis for a claim and aspiration for a sustainable urban future (McDaniel and Lanham 2010) and for the social change needed to arrive at such a future (Holden 2008; Garmendia and Stagl 2010; Reed et al. 2010).

The connection between social learning and democratic deliberation is a close one. At one level, city and regional governments in this era of place competitiveness express new interests in learning from others, to the extent that learning actually derives from participation in international conferences and networks of cities promoting their best practices (Campbell 2009; Seymoar et al. 2009). Advancing a participatory public ideal is key to this image. Public participation gives rise to a binary understanding of participation: are you in or are you out? By creating a deliberative environment that is oriented toward social learning, we invite participants to interact with one another in order not to wave a flag or cast a vote but "to develop an understanding of each other's claims" (Kanra 2007: 4). This involves the skill of "cognitive objectivity," (ibid.: 4), a relational process which evaluates the perspectives of others with regard to the questions confronting the partnership. We seek to identify differences that distinguish and account for the independent views of our fellows and to contrast these with our views.

This understanding of the primary and paramount role of deliberative process is aligned with Hans-Georg Gadamer's (1975) concept of horizon seeking, where the horizon is the elusive common interest. Individuals arrive at a public debate from different perspectives, giving each of them a different vantage point to the horizon. The process of horizon seeking is the discursive work of articulating to others how one sees the horizon, the place where one stands in order to see it that way, and the essential steps in the path to that point. It is emphatically different from the purpose of deliberation which is often asserted in partnerships: that of reaching agreement. This hermeneutic practice involves a view of social learning in which

> understanding always operates with a practical aim in mind and the fusion of horizons works as a test of the claims of each horizon. It is only by this process that individuals come to an understanding of themselves and situate themselves correctly within view of "one great horizon."
>
> (Kanra 2007: 22)

Ostensibly sophisticated learning-oriented initiatives bring the spectrum of interests to specify measures of success, to establish a system and methodology for measuring indicators of sustainability, and to include criteria of success within existing project decision-making structures. These specifications are incorporated by requirement in many city development processes. Studying the unravelling of one such project, in Southwark in London, Rydin et al. (2003: 556) report that they "have had little impact on the decisions that were actually taken." Instead, sustainability considerations, even though embedded in the partnership structure in these specific ways, were reduced to something that was meaningless within the context of the operational decisions that the partnership needed to make. They attribute the failure to learn to "the lack of any agreed consensus on what sustainable development meant in the context of an urban regeneration project" and specifically "little, if any, cultural assimilation of the policy objective of sustainable development" (ibid.: 556). We could draw the same conclusion in order to explain the lack of social justice stipulations, measures and requirements to have an overall impact on urban projects (Holden 2009).

Different from the conventional understanding of "engagement" in community development and social science circles, typically evoking a specific ideal of participatory democracy by empowered and self-actualizing individuals, pragmatic "engagement" is a more encompassing set of activities. It encompasses an individual and place-based authenticity. Still thoroughly integral to the attainment of democracy, engagement in the pragmatic view "indicates a relation to yourself through the environment, in time, and not only to the present situation … engaging with an appropriate environment is a condition for enacting a certain beneficial capacity" (Blokker and Brighenti 2011: 12). That is, effective engagement is a relationship between actors and environments, and understanding that engagement demands attention to both. Both actors and environments are constrained in important ways and cannot be persuaded or compelled by theoretical argument about the benefits of democracy to pursue any given course of engagement. These constraints put different arguments into different regimes of engagement – not strictly boxes that cannot be escaped from but "adherences" in every form of engagement, from the most intimate to the most functional (ibid.: 13).

Social learning emerges within given human lifeworlds of everyday experiences that are framed in societal contexts (Nielsen and Nielsen 2007). This creates different possibilities for transformation when compared to an approach that focuses on social learning through institutions and organizations. In particular, it changes the power dynamics of the engagement scenario. Working in Denmark, Nielsen and Nielsen conducted critical action research into public dialogic processes. They found that typical mainstream dialogic processes undervalue the voices grounded in the lifeworld because these are far too easily overpowered by the established professional discourses. Sufficient time is needed to facilitate the composition of a shared lifeworld through the shared time and shared experiences in the project. The time and experience need to be sufficient to displace the pull of power with the pull of new social ties. An even more basic task is conceptualizing a difference between formal power-based relationships and informal social relationships that are always and already also grounded in power dynamics.

They have designed and convened 'future workshops' which seek to empower actors positioned outside the dominant structures of power, as citizens, employees, or residents. The workshops are directed towards voicing participants' critiques of an existing societal condition that affects their daily lives, via paradoxes and conflicts that they experience, formulating visions for desired changes, and exploring possible pathways for collaborative action to actualize these utopias. The wager here is that selection of participants, roles, and the nature of the engagement can 'void' the dialogue of its imbricated formal power dynamics by filling this void with rich everyday life dynamics. It is a slow and diffusive process of displacement as human relationships build and counter dominant power-based organizational relationships. Social learning is thus richly interpersonal as well as discursive and completely and inseparably immersed in the lives of participants.

The important work lies in locating the cracks and fissures in the existing social structure of everyday life, uncertainty in power dynamics, organizational control, responsibility, and implications. This will enable participants to make good use of engagement opportunities to initiate a transition toward a more sustainable and just future. In Danish, this is referred to as a *Bildung* ideal – promoting reflection and learning on what best serves the common interest, and developing citizen capacity for mutual responsibility (Johansson and Læssøe 2008). Others refer to this as creating a stable urban knowledge arena: a space intended for sustained reflection on the public interest (Nolmark et al. 2009). Nolmark et al. point to four key success factors: (1) increasing mutual or common understanding among participants; (2) improvement of the design of a particular project or policy; (3) engendering new adherents across a wider spectrum of actors; and (4) attainment of the goal that the arena was initially created to meet.

A persistent and pervasive reluctance to learn obstructs this work, particularly in conditions of structured, hierarchical organizations with many diverse actors, diverse and specific job descriptions, and conditions demanding efficiency. This same kind of anti-learning behaviour ultimately bars the path to reflexive learning that questions our beliefs at the same time as it opens us to taking on new beliefs. Rule-following behaviour is based upon history and what can be established based on existing practice; deviating from this kind of behaviour is where forward-looking logic, intentionality, and learning begin to apply (Brunsson and Jacobsson 2000). Within a social learning approach, structures and cultures of learning are key to guarding against the tendencies of "rule-following behaviour." Brunsson and Jacobsson (ibid.: 11) further specify that rule-following behaviour responds to questions such as: "Who am I? What sort of situation am I in? What is appropriate for a person like me to do in a situation like this?" By contrast, social learners acting with a more pragmatic sense of intentionality answer the questions: "What will I like or prefer in the future? What action alternatives are open to me? What would be the future consequences of each of these alternatives?"

Decision-making is quick to usurp all other purposes of policy processes, shifting the horizontal flow of information to a vertical flow intended to lead toward the most expedient and agreeable decision. To counter this, deliberative processes need to be recognized as distinct forms of "reflection-inducing communication that can never be assumed to exist" (Dryzek and Niemeyer 2010: 81) and that will require distinctive time, resources, support, and maintenance efforts. These processes need to be formally and carefully linked to decision-making, too, because

> if citizens cannot link their deliberative practice to the decisions made on their behalf and if there is no clear formula to narrow the gap between citizens and their representatives, then sustaining the level of engagement and deliberative capacity, flourished during the social learning oriented practices of citizens, could become doubtful.
>
> (Kanra 2007: 29)

Empowered space, in the words of Dryzek and Niemeyer (2010: 11), needs to answer to public space. What tangible forms this separation and linking of deliberative and decision-making forums ought to take is an open question.

Others include dialogue platforms dedicated to learning about a particular urban phenomenon, sharing and cataloguing information of common concern, and building a sense of commonality among joiners. Examples include websites dedicated to documenting the loss of particular sites or types of sites in the city, like Jeremiah's Vanishing New York, or Vanishing Vancouver. Others include organizations and websites dedicated to sharing information about and building support for local cultural events, initiatives and businesses: such as Keep Portland Weird or Melbourne's Urban Happiness Facebook discussion group.

One of the most powerful resources deployed in deliberative social learning processes is that of rhetoric. Through alteration of the terms of political discourse, people in dialogue change the understandings of actors of what is and is not worthy, and what is and is not at stake (ibid.). As Habermas (1996: 486) put it, "Communicative power is exercised in the manner of a siege. It influences the premises of judgment and decision making in the political system without intending to conquer the system itself." However, Habermas (1992: 452) also expressed reservations about relying on the power of discursive communication alone to change understandings and actions, concluding that "communicative power cannot supply a substitute for the systematic inner logic of public bureaucracies." Communicative power suffers from a lack of structure that can be grounds for dismissal of its value by organizational structures. Dryzek and Niemeyer (2010: 33) argued "that the very conditions of structurelessness favored by deliberative democrats are exactly the conditions most conducive to arbitrariness, instability, and so manipulation in collective choice." Habermas (1996: 373) agreed that "the signals they send out and the impulses they give are generally too weak to initiate learning processes or redirect decision-making in the political system in the short run."

Of course learning and knowledge-generating processes are subject to power dynamics and manipulation. They are social processes. But this is never all that is going on. Interactions of knowledge, and contextual understanding and power in all its forms operate in complex ways "that cannot be reduced [solely] to manipulation strategies" (Thévenot 2007: 414). Through a "civilizing force of hypocrisy" (Elster 1998: 12), participants in this kind of negotiation are sometimes able to persuade themselves of the validity of the perspective they are advancing, even if they were not so sure they meant it at the outset. This experience of persuasion can, in Thévenot's terms, raise the generality of a claim or justification to become more encompassing of what others would experience as good and true. In John Forester's (1999: 230) justification of talk as action, "what gets done depends heavily on what gets said."

Formalization and insistence upon the need and the role for social learning as a first step in deliberative practice are key to a long-term strategy to infiltrate learning into the inner logic of institutions, be they public or private. The critique of structurelessness has force so long as deliberation is a prelude to aggregation of opinion, usually by voting based on majority-rules principles. However, if we reconceptualize public deliberation in terms of providing outcomes of language, learning, and the contestation of discourses as transmitted to the state, this critique dissolves. The notion is to create the conditions for deliberation as a separate but institutionalized component of accountability of partnerships to the public, such that the possible justifications for decisions and actions are discussed in public before the decisions and actions themselves are determined (Dryzek and Niemeyer 2010).

In the broadest sense, the social learning that is needed to create the conditions for the Great Community will occur in the interaction of knowledge with governance arrangements toward new understandings of urban practices and public goals (Lee 1993; Parson and Clark 1995; Meppen and Gill 1998; Bulkeley 2006; Stagl 2007; Atkinson et al. 2011). This is true at a number of levels: cities recognizing the need for technical know-how regarding the practices and policies of urban development; the need to better track, apply, and make public key information on significant local trends; to generate replicable knowledge from practice; to improve avenues for city council and staff related to peer learning, exchange of ideas, and cooperative work at home and internationally; and to create effective channels for two-way learning with the community at large.

## Pragmatic reappropriation of empowerment in community

Without being surprised by or ignorant of the sometimes insidious dynamic of empowerment discussed in Chapter 3, the blinders of individualistic authenticity described in Chapter 4, or the lack of association of learning with power in many contexts, the new generation of urbanism can find hope in an approach to engagement that does not take itself nearly as seriously as utopianisms of the past. These new ideals are expressed not in a Garden City-like ideal, but may in fact represent a closer approximation of the marriage of vision and science that Mumford called for in his story of utopias. They take the form of neighbourhood-scaled smart growth and transit-oriented development strategies, green building standards, commuter cycling challenges, car sharing and all manner of mobile technology-facilitated new cooperatives. They are community-based investment circles, that bring communities together with small contributions and common dreams of launching and building community-owned cooperative businesses or cohousing projects that bring communities together to invest their ideas, skills, and labour as well as capital in order to build a common home (Figure 5.3).

These numerous innovations are born from necessity, or at least practicality, as testament to whatever vision they also represent. If they move us toward sustainability and justice goals at all, it is only piecemeal, incompletely. They move us sideways and backwards from these goals as well. They are processes of 'learning by doing,' investments of energy and hope in the prospect of innovation originating in communities (Mendell and Neamtan 2010: 71). They are sometimes very small-scale expressions, with stronger doses of individual authenticity than of community engagement and empowerment (Figures 5.4–5.7).

These grassroots mobilizations do not always fit the expectations of classical social movement frames. In particular, the insistence upon a regime of

*Figure 5.3* Cohousing participatory construction site, Strasbourg, France. Citizen-led
        participatory housing construction is encouraged by the municipality on
        selected available infill development sites in French cities such as Strasbourg.
Source: Photo, the author.

engagement oriented around individualism as opposed to the collective poses
problems when it comes to articulating concerns and values persuasively to
others. (Thévenot does point out that even union activists, within a classical
social movement framing, sometimes make use of individualized argument
drawing upon authenticity when engaging with other workers and activists.)
Personal passions need frequently to be re-"formatted" in order to integrate
with common interests. Sometimes this reformatting happens in a transformative way, as when an individual speaking up to protect a charismatic old tree
within view of her kitchen window comes, through communication, learning,
and engagement, to transform her argument. Where it may have once centred
on personal attachment and privilege, the argument might transform to one of
the ecological, habitat, and ecosystem services value of large trees, justice in the
distribution of trees as well as the important stormwater and temperature regulation roles that trees play. In the critically charged environment of cities, even

*Figure 5.4* Development permitting dispute on a development application notice board, Mount Pleasant, Vancouver. On this development permit application notice board, posted as per city regulation on the property to which the application applies, a written dispute is inscribed in black marker. The first disputant has drawn a square around the notation of an intended "3 car garage" in the description of the proposed development, and drawn a line to append the comment: "Housing for rich people!" A second disputant has written adjacent to this comment: "Who contribute & PAY TAXES!" Another disputant, or perhaps the first disputant again, has written back: "Right because everyone who works can afford to live here." A fourth comment concludes the argument: "And you're proud of paying taxes to this corrupt government?" Of note, perhaps missed by the first disputant, is that the three-car garage plus one surface parking space in the rear of the proposed development are to serve four dwelling units proposed for this site.
Source: Photo, the author.

urban experimental failures in this kind of empowered engagement can add to the texture and conflict that incline us toward understanding a better, uniquely urban, future. They could represent, in Cruikshank's (1999) terms, a reappropriation of empowerment.

The shift in discourse represented by the pro-urban political idea that cities do important work as economic and political innovators is empowering and motivating in this context, so long as notions of 'innovation' remain scaled ultra locally. This is not for any presumed virtue of this geographical scale – we do not want to fall victim to the local trap discussed in Chapter 3. It is, rather, for the sake of dissuading the force of capitalist growth motivations from holding sway over other, social motivations. While power tends to concentrate and

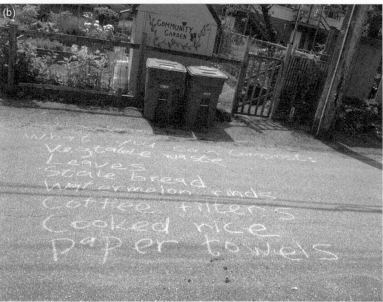

*Figure 5.5* (a) and (b) What you can and can't compost, Mount Pleasant neighbourhood, Vancouver. Simple, pop-up street intervention in social learning and behaviour change, by an individual aiming to provide a logic and justification for neighbours to change their waste disposal habits.

Source: Photo, Omer Rashman.

*Figure 5.6* Will you @CatchEveryDrop? Vancouver, BC. Another sidewalk chalk effort, paired with a Twitter campaign, bringing summer water restrictions to the neighbourhood scale. In addition to raising awareness about water conservation, this commoning work challenges the private property rights perspective toward the green lawn through a neighbourly intervention calling this greenness into the realm of common property.

Source: Photo, Kate Elliott.

centralize, participation remains local. Important decisions are made in board rooms and city council chambers; but learning happens elsewhere (sometimes at the bar after the meeting, sometimes in a cold sweat in bed later that night). Orienting our understanding of this dynamic of learning and decision-making, whereby the "ayes" are counted in board rooms and city hall chambers, but the "eurekas" are had in parks, garages, coffee shops, and street corners, is key. This way, people can persist in their belief that what they may achieve working in their own neighbourhoods and cities may matter and make a difference. People continue to be willing to commit at the local level, and neighbourhoods are small enough to resist the corrupting influence of capital on scalable work.

If the vision can be kept in flux and not fixed or idealized, it will retain credibility even among the radically disillusioned. The new urban utopias can be put in practice in steps that can be taken from anyone's starting point or level of consciousness. In this way, an urban garden can be just a garden; it is not going to change the global food import-export balance or feed the city. A car-free day is not going to persuade the city to abandon all personal automotive transportation for all time. A critical mass bicycle ride might not be a serious protest against cars' occupation of the roadway, it might just be a fun thing to do with friends. While such activities may then seem vain, idiosyncratic, or

*Figure 5.7* Plant food not lawns. Abbotsford, BC. An attempt to advance a justification
of thrift and humility, along with a campaign for gardening and exchanging
seeds and plants.
Source: Photo, Tammas Grogan.

ridiculous, they are movements outside of prevailing dominant visions of life in
cities, and thus represent tentative steps in a new urban engagement and
empowerment. But whatever they may represent in theory, they are no more
than enjoyable in practice; and whatever they represent in distance from the
mainstream, they do not make sense as overarching ideals. In experiments with
these activities over time may lie the potential to remake ourselves into more
pragmatically ethical actors. Referring to the specific power of citizen efforts in
"commoning" urban spaces, or creating commonly held spaces out of public
spaces, Chatterton (2010: 626) suggests that, as cynical as we may be about this
work, there is always more going on in the subtext of social behaviours than
first meets the eye:

> We should not position the common as something always subjugated or in
> response to the more dynamic practices of capital accumulation. The

common is full of productive moments of resistance that create new vocabularies, solidarities, social and spatial practices and relations and repertoires of resistance.

Coming to view these acts of urban engagement and empowerment as both radical and insignificant may be an essential skill for individuals trying to make sense of their authentic self and agency. Those who master it may find more to value in a larger human and more-than-human urban community, in a frighteningly overdetermined urban world.

## Conclusion

We are faced with both limitations and constructive opportunities to support a more pragmatic orientation toward justice and sustainability in the city. Our understanding of 'community' has been badly damaged, compared to historical aspirations for a Great Society, but the prospects for the Great Community are not nil. The case for public space continues to be made loud and clear, but there is reason to doubt whether our planning for and maintenance of truly open-minded space is on steady ground. Looking at the new design frontier in the actual manifestation of urban placemade neighbourhoods, explicitly seeking to serve wellbeing, we can apply an understanding of these limitations and opportunities to the improvement of the outcomes in urban practice.

Context does matter. The positions that people take in a dispute depend upon the context in which they perceive themselves, which depends upon the objects they align themselves with, the path they take, and the material worlds they see occupied by others and that they engage. In a complicated world, what we do may or may not be consistent with the beliefs we purport to uphold. Therefore, in any given dispute, it makes more sense to study the discursive and material actions taken than to position people *prima facie* according to one ideological camp or the other. Different situations present different alternatives to an individual, making one or other competing discourse more appealing, more available, and more resonant. At the same time as structure and context mean everything to the outcome of a negotiation or dispute, a pragmatic view holds that each situation is also radically open to emergent, even transcendent understandings that may arise through the pursuit of new social learning. Such an outcome is always contingent.

Process matters too. For pragmatist Hans Joas, pragmatism aims toward the perfection of process via an attempt "to reconstruct empirically the procedure for resolving situations requiring moral decisions in such a way that in it a procedure becomes recognizable that can itself serve as a basis for its own self-perfecting" (Joas 1993: 252–253).

One test of the potential for justice within urbanity today, is, therefore, a test of alignment: are our collective works, at home in our cities, producing a material reality that is more closely aligned with our diverse, individualized moral purposes, and intentions of empowered participation? The test of successful engagement processes could be whether we are creating conditions for people to work together in order to better walk their own talk. This would constitute urban planning in sum as a kind of "situated moral inquiry … aimed at formulating a principle of justice to guide a solution rather than importing a universal principle of justice to apply in an act of final moral judgment" (Lake 2016). Taking a critical pragmatic approach demands this commitment to collective knowledge generation, the process of arriving together at these principles in particular situations.

At the turn of this century, "the commons" was considered to be a vestige of an earlier time and system of social organization that was "destined to disappear in the face of modernization" (Agrawal 2002: 42). Graham and Marvin (2001) set the tone for the thinking of the day, namely, that globalized capitalism in the form of neoliberalism was grossly intensifying the fragmentation of urban landscapes into privatized enclosures of structures, infrastructures, and services. The ultimate conclusion of this process would be the splintering of all common spaces and perhaps all human groups, too, into fragments of their former collectivities, each dedicated to specific individualized purposes. More recently, a renewed interest in the continuing relevance of "the commons" has arisen. This interest began with Elinor Ostrom's (1990) definitive work on common property resources and has moved on to consider "the practice of commoning" within our cities (Chatterton 2010: 626). In Chapter 6, I deepen the investigation of the potential role for a pragmatic approach to urban sustainability and justice by taking up the question of the commons within the consideration of urban risk and resilience.

# References

Aboelata, Manal, Leah Ersoylu, and Larry Cohen. "Community Engagement in Design and Planning." In *Making Healthy Places: Designing and Building for Health, Well-being, and Sustainability*, edited by Andrew Dannenberg, Howard Frumkin, and Richard Jackson. Washington, DC: Island Press, 2011, pp. 287–302.

Agrawal, Arun. "Common Resources and Institutional Sustainability." In *The Drama of the Commons*, edited by Elinor Ostrom. Washington, DC: National Academy Press, 2002, pp. 41–86.

Atkinson, Robert, George Terizakis, and Karsten Zimmerman. "Introduction: Governance, Knowledge and Sustainability – An Introduction and Overview." In *Sustainability in European Environmental Policy: Challenges of Governance and Knowledge*, edited by Robert Atkinson, George Terizakis, and Karsten Zimmerman. London: Routledge, 2011, pp. 1–10.

Beauregard, Robert A. *Planning Matter: Acting with Things*. Chicago: University of Chicago Press, 2015.

Bell, Simon and Stephen Morse. "Breaking Through the Glass Ceiling: Who Really Cares about Sustainability Indicators?" *Local Environment* 6(3) (2001): 291–309.

Blokker, Paul and Andrea Brighenti. "An Interview with Laurent Thévenot: On Engagement, Critique, Commonality, and Power." *European Journal of Social Theory* 14(3) (2011): 383–400.

Boltanski, Luc and Eve Chiapello. *The New Spirit of Capitalism*. London: Verso, 2005 [1999].

Boltanski, Luc and Laurent Thévenot. *On Justification: Economies of Worth*. Princeton, NJ: Princeton University Press, 2006 [1991].

Brunsson, Nils and Bengt Jacobsson. *A World of Standards*. Oxford: Oxford University Press, 2000.

Bulkeley, Harriet. "Urban Sustainability: Learning From Best Practices?" *Environment and Planning A* 38 (2006): 1029–1044.

Campbell, Tim. "Learning Cities." *Habitat International* 33(2) (2009): 195–201.

Chatterton, Paul. "Seeking the Urban Common." *City* 14(6) (2010): 625–628.

Cowell, Robert and Susan Owens. "Governing Space: Planning Reform and the Politics of Sustainability." *Environment and Planning C: Government and Policy* 24 (2006): 403–421.

Cruikshank, Barbara. *The Will to Empower: Democratic Citizens and Other Subjects*. Ithaca, NY: Cornell University Press, 1999.

Dewey, John. "Search for the Great Community." In *The Philosophy of John Dewey*, edited by J. J. McDermott. Chicago: University of Chicago Press, 1981 [1927], pp. 620–643.

Dewey, John. *The Public and its Problems*. New York: Henry Holt & Co., 1927.

Dryzek, John and Simon Niemeyer. *Foundations and Frontiers of Deliberative Governance*. Oxford: Oxford University Press, 2010.

Eicher, Caitlin and Ichiro Kawachi. "Social Capital and Community Design." In *Making Healthy Places: Designing and Building for Health, Well-being, and Sustainability*, edited by Andrew Dannenberg, Howard Frumkin, and Richard Jackson. Washington, DC: Island Press, 2011.

Elster, Jon. *Deliberative Democracy*. London: Cambridge University Press, 1998.

False Creek Development Group. *Inner City Living: The False Creek Redevelopment Program: A Demonstration Project under the Canadian Urban Demonstration Program*. Vancouver: City of Vancouver, 1974.

Fishman, Robert. *Urban Utopias in the Twentieth Century: Ebenezer Howard, Frank Lloyd Wright, Le Corbusier*. Cambridge, MA: MIT Press, 1982.

Forester, John. *The Deliberative Practitioner: Encouraging Participatory Planning Processes*. Cambridge, MA: MIT Press, 1999.

Friedmann, John. *The Prospect of Cities*. Minneapolis, MN: University of Minnesota Press, 2002.

Fung, Archon and Eric Wright. *Deepening Democracy: Institutional Innovations in Empowered Participatory Governance*. New York: Verso, 2003.

Gadamer, Hans G. *Truth and Method*. New York: Seabury, 1975.

Garmendia, Eneko and Sigrid Stagl. "Public Participation for Sustainabilty and Social Learning: Concepts and Lessons from Three Case Studies in Europe." *Ecological Economics* 69(8) (2010): 1712–1722.

Graham, Stephen and Simon Marvin. *Splintering Urbanism: Networked Infrastructures, Technological Mobilities and the Urban Condition*. New York: Routledge, 2001.

Habermas, Jürgen. *The Theory of Communicative Action*. Trans. T. McCarthy. Boston: Beacon Press, 1984.

Habermas, Jürgen. "Further Reflections on the Public Sphere." In *Habermas and the Public Sphere*, edited by Craig Calhoun. Cambridge, MA: MIT Press, 1992, pp. 421–461.

Habermas, Jürgen. "Three Normative Models of Democracy." In *Democracy and Deference*, edited by Seyla Benhabib. Princeton, NJ: Princeton University Press, 1996, pp. 21–30.

Holden, Meg. "Social Learning in Planning: Seattle's Sustainable Development Codebooks." *Progress in Planning* 69 (2008): 1–40.

Holden, Meg. "Community Interests and Indicator System Success." *Social Indicators Research* 92(3) (2009): 429–448.

Holden, Meg. "Public Participation and Local Sustainability: Questioning a Common Agenda in Urban Governance." *International Journal of Urban and Regional Research* 35 (2) (2011): 312–329.

Joas, Hans. *Pragmatism and Social Theory.* Chicago: University of Chicago Press, 1993.

Johansson, Magnus and Jeppe Læssøe. "Mediator Competencies and Approaches to Participatory Education for Sustainable Development." Paper presented to AERA, New York, March 24–28, 2008.

Kanra, Basra. "Binary Deliberation: The Role of Social Learning and the Theory and Practice of Deliberative Democracy." Joint Sessions. Helsinki:European Consortium for Political Research, 2007.

Katz, Amy, Rebecca Cheff, and Patricia O'Campo. "Bringing Stakeholders Together for Urban Health Equity: Hallmarks of a Compromised Process." *International Journal for Equity in Health*, 14(1) (2015): 1–8.

Lake, Robert W. "Justice as Subject and Object of Planning." *International Journal of Urban and Regional Research* (2016). In press.

Lee, Kai. *Compass and Gyroscope: Integrating Science and Politics for the Environment.* Washington, DC: Island Press, 1993.

Lippmann, Walter. *Public Opinion.* New York: Harcourt, Brace, 1922.

Lynch, Kevin. *The Image of the City.* Cambridge, MA: MIT Press, 1960.

Lynch, Kevin. *A Theory of Good City Form.* Cambridge, MA: MIT Press, 1981.

McDaniel, Reuben J. and Holly J. Lanham. "Sustainable Development: Complexity and the Problem of Balance." In *Pragmatic Sustainability: Theoretical and Practical Tools,* edited by Steven A. Moore. New York: Routledge, 2010, pp. 51–66.

Mendell, Marguerite and Nancy Neamtan. "The Social Economy in Quebec: Towards a New Political Economy." In *Researching the Social Economy*, edited by Laurie Mook, Jack Quarter, and Sherida Ryan. Toronto: University of Toronto Press, 2010, pp. 87–115.

Meppen, Tony and Roderic Gill. "Planning for Sustainability as a Learning Concept." *Ecological Economics* 26: 121–137, 1998.

Michels, Ank and Laurens de Graaf. "Examining Citizen Participation: Local Participatory Policy Making and Democracy." *Local Government Studies* 36(4) (2010): 477–491.

Mouffe, Chantal. *The Democratic Paradox.* New York: Verso, 2000.

Mumford, Louis. *The Story of Utopias.* New York: Boni and Liveright, 1922.

Nielsen, Kurt A., and Birger S. Nielsen. *Demokrati og Naturbeskyttelse: Dannelse Af Borgerfællesskaber Igennem Social Læring – Med Møn Som Eksempel.* [Democracy and Nature Protection: Forming of Citizen Communities Through Social Learning – Exemplified by a Process Run at Moen, Denmark]. Copenhagen: Frydenlund Academic, 2007.

Nolmark, Henrik, Hans T. Andersen, Rob Atkinson, *et al.* "Urban Knowledge Arenas: Rethinking Urban Knowledge and Innovation." Final Report of COST Action C20. Gothenburg, Sweden, 2009. Available at: www.unige.ch/ecohum/Associations/C20_Final_Report.pdf (accessed 18 August 2016).

Ostrom, Elinor. *Governing the Commons: The Evolution of Institutions for Collective Action.* Cambridge: Cambridge University Press, 1990.

Parson, Edward A. and William C. Clark. "Sustainable Development as Social Learning: Theoretical Perspectives and Practical Challenges for the Design of a Research

Program." In *Barriers and Bridges to the Renewal of Ecosystems and Institutions*, edited by Lance H. Gunderson, C. S. Holling, and Stephen S. Light. New York: Columbia University Press, 1995, pp. 428–460.

Quarter, Jack and Laurie Mook. "An Interactive View of the Social Economy." *ANSERJ* 1(1) (2010): 8–22.

Raco, Mike. "Spatial Policy, Sustainability and State Restructuring: A Reassessment of Sustainable Community." In *The Sustainable Development Paradox*, edited by Rob Krueger and David Gibbs. London: Guilford Press, 2007, pp. 214–238.

Reed, Mark, Anna Evely, Georgina Cundill, *et al.* "What is Social Learning?" *Ecology and Society* 15(4) (2010): r1.

Rose, Nikolas. "The Death of the Social? Re-Figuring the Territory of Government." *Economy and Society* 25(3) (1996): 327–356.

Rydin, Yvonne. *Governing for Sustainable Urban Development*. London: Earthscan, 2012.

Rydin, Yvonne, Nancy Holman, Vicky Hands and Florian Sommer. "Incorporating Sustainable Development Concerns into an Urban Regeneration Project: How Politics Can Defeat Procedures." *Journal of Environmental Planning and Management* 46 (4) (2003): 545–561.

Seymoar, Nola-Kate, Zoe Mullard and Marena Winstanley. *City-to-City Learning*. Vancouver: Sustainable Cities International, 2009.

Stagl, Sigrid. "Theoretical Foundations of Learning Processes for Sustainable Development." *International Journal of Sustainable Development and World Ecology* 14 (2007): 52–62.

Thévenot, Laurent. "The Plurality of Cognitive Formats and Engagements: Moving Between the Familiar and the Public." *European Journal of Social Theory* 10(3) (2007): 413–427.

Thévenot, Laurent, Michael Moody and Claudette Lafaye. "Forms of Valuing Nature: Arguments and Modes of Justification in French and American Environmental Disputes." In *Rethinking Comparative Cultural Sociology: Repertoires of Evaluation in France and the United States*, edited by Michèle Lamont and Laurent Thévenot. New York: Cambridge University Press, 2000, pp. 229–272.

Thomas, Marlo. "Free To Be You and Me." Audio Recording. Bell Records, 1972.

United Nations. *Rio Declaration on Environment and Development*. Rio de Janeiro, Brazil: United Nations, 1992. Available at: www.unep.org/documents.multilingual/default. asp?documentid=78&articleid=1163 (accessed 18 August 2016).

United Nations. "Local Government Declaration to the World Summit on Sustainable Development. Johannesburg, South Africa." 2002. Available at: www.stakeholder forum.org/fileadmin/files/AMR_2008/LocalGovtDeclaration.pdf (accessed 18 August 2016).

Walzer, Michael. "Pleasures and Costs of Urbanity." In *Metropolis: Center and Symbol of Our Times*, edited by Philip Kasinitz. New York: New York University Press, 1986, pp. 320–330.

World Health Organization and United Nations Human Settlements Programme. *Hidden Cities: Unmasking and Overcoming Health Inequities in Urban Settings*. Kobe, Japan: World Health Organization and UN-Habitat, 2010.

Yanarella, Ernest J. "Local Sustainability Programmes in Comparative Perspective: Canada and the USA." *Local Environment* 4(2) (1999): 209–223.

Zarate, Loreno. "Right to the City for All: A Manifesto for Social Justice in an Urban Century." In *The Just City Essays: 26 Visions for Urban Equity, Inclusion and Opportunity*, edited by Toni Griffin, Ariella Cohen, and David Maddox. V.1. New York: J. Max Bond Center on Design, Spitzer School of Architecture, City College of New York, Next City and The Nature of Cities, 2015, pp. 26–29.

# 6   Security, struggle, and resilience in the city

Key to the contemporary celebration of the city is the sense of rarefied order and security that parts of the placemade city now offer. The city now offers a baseline sense of safety, crucial to a willingness of many to participate in the public sphere. And the city is also the nucleus of what Ulrich Beck (1992) called the "risk society," a society in which uninsurable, unavoidable, intrinsic risk features as a key characteristic, in need of management. Living in a globalized risk society, as per Beck, means that all people are subject to unprecedented global risk. This is the basis of the demand for "the guaranteed city." Indeed, the risk that we are aware of confronting every day is presented as integral to, auto-genic to capitalism itself. The risk sits within all the "disaster-bearing structures" of modern society, a matter "of unprecedented endangerment," going all the way down to our identities, threatening the agonies of "self-dissolution, self-endangerment and self-transformation" (Beck 2009: 163; Gleeson 2015).

The drive to counter chaos and risk with order is a central and long-standing one in urban politics, of both street-level and formal varieties. The modern ideal strives to create conditions of equality among people. A fundamental role of the modern nation state is to ensure people's bodily security and security of their property, regardless of their position or wealth. At the ground level of the city street, equality under conditions of diversity is considered to be best served by non-confrontation. So cities aim to deliver a pacified and docile public space, capable of being non-offensive and accessible to all. They rely on a culture of civility as the means to arrive at order and security. This sense of value is also compatible with practices of enclosure and privatization of public meeting places, all manner of barriers and walls, reliance upon automated or visible physical security presence in public places. It is also compatible with enforcement to clear areas of people judged not to belong and to pose a risk to the order in humiliating and undignified ways. All of these practices constitute the structures of the social organization of risk. Adorno (2002 [1941]: 445) calls the impact of this work on the urban population not civility but: "pseudo-individualization … endowing cultural mass production with the halo of free choice or open market on the basis of standardization itself, which keeps customers 'in line' by means of 'pre-digested' products." Indeed, civilized and docile public spaces can create and reinforce a strong sense of segregation.

Against this demand for a sense of safety in the city is an opposite demand, for what is perhaps an even more enduring urban quality. Risk. The risk within contemporary society threatens all that we know about ourselves and about life itself. It exemplifies "self-endangerment." At the same time, within this risk is also the possibility and prospect for "self renewal, of the basic institutions of the nation-state and industrial society" (Beck 2009: 163). Risk is the allure, the excitement, the very prospect of transformation in the city.

Safety and risk clearly contradict one another. The rise of the first can kill the prospect of risk and chance that is crucial to the promise of the city as a site of unprecedented pursuit of sustainability and justice. At the same time, battling and quelling risk entail a threat to the sense of urban authenticity, urban empowerment, and urban community discussed in Chapters 4 and 5. There is a strong sense within the critique of neoliberalism that the rolling out of these systems of securitized governance restrict public spaces and the work of communities, and force self-governance, self-regulation, and the regulation of relationships between individuals (Foucault 1991). It is also pointed out that this form of securitized, guaranteed city is the antithesis of a learning-oriented city, discussed in Chapter 5. Along these lines, Beck (2009: 115; also Gorz 2010) refers to the notion of the "knowledge society" today as a euphemism for its very opposite. In his view, the risk society "is a non-knowledge society in a very precise sense." Instead, the maintenance of defence against pervasive risk takes on different forms of non-knowledge, including "selective assumptions, wilful ignorance, reflected non-knowing, conscious inability to know, unconscious non-knowing and the inability to know" (Beck 2009: 126). In response to the risk society, humanity "has marched into the age of unknowing, poised between 'ambivalences of more-modernity' and 'ambivalences of anti-modernity'" (ibid.: 231; Gleeson 2015: 9). To the extent that urbanites are threatened by risks we can scarcely predict or guard against, we are incapable of achieving our life plans, of empowering our voices and actions, of feeling at home in the common place, of learning openly with others. And yet, when we feel that our lives, loved ones, homes, and communities are in peril, all those sacrifices seem worthwhile.

In this chapter, we consider the tension between risk and safety in the contemporary city, and the bridging concept of urban resilience as key to urban sustainability and justice. Forces in the city that push toward securitization against risk create more gates and barriers, more "safe" public spaces along with more of their opposite: marginal spaces in which to contain marginalized people pushed out of the securitized spaces in the name of safety. These forces call the sustainability of a common public square into question. In many cities, these forces increase the privatization of urban space, leading to critiques of the extreme commodification of the land and a concomitant destruction of its common nature.

## Urban risk and urban reward

Brugmann's (2009: 161) explanation of cities' renaissance is anti-utopian: "they are an honest, expedient response to a great demand." But the city is not only expedient, not just a response; it is a reward, an event to be celebrated, something beloved. It is a destination experience, not just an instrument of convenience. It is now an essential, dominant human experience. The reward offered by the urban experience today is to be free to pursue whatever specialized pursuit one may choose and excel at; to advance individual freedoms and to feel personally empowered in so doing. In many cases, this freedom rights inequities that would otherwise keep many confined to menial labour. Of course, this is not true in all cases. Many people are stuck doing menial labour, anyway, and it is still fairly predictable what these people will look like, which part of town they live in, what kind of education they never had a chance to receive. Geographic, income, and ethnic determinations aside, there remain many other downsides of this urban reward. Significantly, all the life choices, as well as careers, divide the city up into tiny urban specializations, resulting in a city full of individuals and small groups, the majority of whom cannot communicate with one another.

Cities represent a risk that many of us refuse to try to insure out of our lives; a risk that makes us feel alive. This kind of urban life runs against the grain of the promise of the "guaranteed city" where legitimate and carefully delimited objectives are monitored by experts and well-framed participatory procedures (Breviglieri 2013). While politicians and planners turn their attention to models and strategies for safety and security, the urban experience retains a vote against the sense of future-proofing. But where is the sweet spot for urban drama, conflict, and excitement? A certain density, a certain rent level, a certain size or age? Pattaroni and Pedrazzini (2010) examine how difference in the urban context is made into the substrate of fear, such that an "urbanism of fear" develops in which policies and practices to combat insecurity become the order of the day. From this politics is generated a large market-driven insecurity industry, restructuring and redesigning the city's public spaces for surveillance and policing, and a built environment of risk aversion. In theory, individual freedoms and empowerment are undiminished, at least for the law-abiding. It would be a mistake, however, to think that this appeal of the city can be divorced entirely from the way in which urban life is experientially risky. A serious side effect is a validation of the destruction of the common good, and urban cohesion. The ideal of the urban public which is diverse and exists in a space of diversity is lost.

In the current wave of urbanism we find a defensiveness toward density, diversity, and size as at least incomplete predictors of urban dis-ease and anomie, and sometimes as offering greater potential for authentic community building than smaller settlements. At the heart of this are diversity and nuance of form, feel, and function of our urban landscapes; something more complicated than a product, brand, or density. Richard Sennett (1970), who offered the quintessential definition of a city as a place "where strangers are likely to meet" (Sennett 1978: 29), has long argued that the density and diversity of urban public space are key to finding urban solutions to social problems. Encounters with diversity in close quarters civilize us, in real ways that contribute to remaking our very identities and that work in the direction of tolerance. At the same time, articulations and celebrations of diversity are detached from the broader systems of power that define cities.

Pursuing an urban life, nonetheless, offers rich experiential reward, which is part of the reason that individual authenticity has caught hold so strongly as an urban way of life today. Urban life is, in this same experiential way, risky. Urbanists like Robinson (2006) and Watson (2006) remind us of the risk that is always lingering on the fringes of otherwise perceptibly safe urban situations, and the anxiety and fear that come along with this dynamic. As such, paying attention to the kinds of cultural and collective expressions that can be found in everyday encounters in urban public spaces is a way of recognizing the deep-seated urge that individuals have to experience "the other." Here, spending time in public space and being open to these encounters, urbanites fight back against the risk society and make "a kind of public commitment to the margin" (Amin 2008: 16).

With a reflexive modernity in which a sense of individualism, for its authentic potential and its loneliness, has eclipsed a sense of class-consciousness, urbanites instead have a taste for taking risks that promise a differentially individualized reward. The risk that is entailed in coming into close contact with people from radically different backgrounds, perspectives, lifestyles and socioeconomic levels than ourselves is a key attraction of the modern city. The stories that we tell about why our cities hold such excitement and stimulation – the risk of coming face to face with those who have it all, those who have something unfathomable to us, and those who have nothing – forces us to confront our own failures and fortunes, weak spots and ignorance, and in some cases, to face threats to our property and bodily integrity as a result of these encounters. This is a uniquely urban risk and reward in the modern capitalist city. And passions have been roused by evidence of the force of just these contradictions of diverse humanity.

The diversity that urban placemaking pursues does not often extend effectively to the poor. Take, for example, the case of the Cape Town Partnership, a central city improvement effort which includes "increasing residential density and access to affordable housing, enhancing public access and mobility to and within the central city, and supporting social development" (Hamann and April 2013: 18). When the partnership's "zero tolerance" policy toward street misdemeanours was shown to be leading to a criminalization of the homeless, the board instituted a new social development programme with a "fundamentally different approach to the homeless" (ibid.: 19). The different approach was street workers who help the homeless to find a home, or "help by paying the bus fare" out of Cape Town (ibid.: 19). This is the Cape Town Partnership's answer to social development, in a country with one of the world's highest levels of socioeconomic inequality, in a metropolitan area in which almost 40 percent of households are estimated to live below the poverty line. In my own city, Vancouver, a private walking tour company called ToursByLocals has recently made news headlines for offering walking tours of the city's poorest neighbourhood for a price from $185–$275. The neighbourhood is in the throes of a heated battle of gentrification and displacement of the marginalized; the tour organizer suggests that "A lot of people will avoid [the neighbourhood]. Instead, I'm showing people the positive aspects and how concepts like social enterprise can make a change and do good" (Puri 2016).

Marginalized people are either seen as pitiful and oppressed, drains on public resources, or else as unfortunate by-products of the necessary course of development – a sad but plausible market for urban entrepreneurs working to monetize and generate new forms of profit at these very margins. As such, within the dominant urban ideal, we lack a strong cultural argument against gentrification, even as gentrification processes become a dominant mode and justification of urban development worldwide. The 'super' or 'hyper' gentrification currently running amok in top-ranking global cities includes a gentrification of critique: not only the poor, but truly public and non-monetized spaces have been pushed out, and so has the kind of thinking that holds that this situation is unacceptable, or even avoidable. Inequality is considered to come with the territory of contemporary urban success. However, if we lose the cultural thread of critique of inequality, we lose the battle of the risk and reward of the city. As forces conspire to destroy the risk and reward of cities, the city also inoculates itself against the potential to produce forces of discord essential to engaging with diversity humanely and democratically.

In this situation, we can see within some quarters of the new urban thinking a new defence mounting for the value of diversity and equity. Certainly, there is a sense among urbanists like Friedmann (2002) and Merrifield (2014), and even occasionally with Richard Florida (2002), that gentrification is a battle against the risk and potential reward of the inclusive city. We scrub our cities clean of that risk and associated reward to the peril of cities themselves. At the same time, recognizing this demand for inclusion is not the end of the challenge. The new urban thinking must also defend against this valuing of risk being appropriated into a tokenistic, oppressive gesture. Even as it critiques displacement in the name of risk and reward, it can itself become a new form of exclusion. The poor of the city become important, as a backdrop, an aesthetic contrast. Streets in marginalized neighbourhoods become the sites of claims of bravery as pioneers move in, invest, authenticate themselves. As a move to more adequately value the risk and reward that inhere to urban life, we need a more critical edge to our thinking.

## The prospect of resilience

The concept of resilience offers a promising means of finding a way through the urban demand for both risk and safety. When applied to cities, resilience thinking promises to reorient the work of planning in a more pragmatic than abstractly ideal direction; that is, more in response to likely future risks and rewards that are recognized and valued in a given context. Part of the focus of resilience planning is to grow and develop social adaptive capacity (Adger 2003) in community to feel empowered and able to prepare and respond in the face of risk (Norris et al. 2008). In *The New York Times* opinion pages, Zolli (2012) offers three reasons for considering resilience to hold the key to better urban planning in our time. One reason is the way resilience thinking centres planners' focus upon the need to respond to environmental threats, rather than reduce environmental impacts. In this way, resilience offers a psychological stretching of the time and alternative possibilities horizon of planning further that is empowering and constructive. Second, resilience planning opens up a connection within the practice of urban planning to human psychological resilience and needs in the face of change. This stretches the scope of technical expertise of planning into adding new domains of better understanding human individuals' aptitude for risk and need for a sense of safety. Finally, Zolli sees potential within resilience for emphasizing the value of more and better data, including social and psychological data, to planning and managing for resilience. Sociologist Andrew Ross (2011: 16) explains that the key areas in need of resilience, in the context of responding to climate change, are "our social relationships, cultural beliefs, and political customs." Individual, social, and political-institutional system change are crucial to our ability to cope with the increasing rate and level of stress and shock promised by an unstable urban social and environmental system. Action in this realm is more crucial than any change to our relationships with the nonhuman world, including greenhouse gas emissions per se. A sense of resilience can help us to focus on this need for adaptive social capacity in all domains of urban life.

The prospect of urban resilience offering a step toward sustainability and justice depends upon the extent to which resilience provides a sufficiently appealing narrative of social, economic, and cultural progress, to get us around the urban stalemate of both desiring and needing to avoid risk. In pragmatic terms, it depends upon the ability of resilience to sit in as a determinant of worth and value. A 2012 "interface" debate on the topic of resilience in the journal *Planning Theory & Practice* (Davoudi et al. 2012) became the journal's most downloaded paper, with a blockbusting 23,000 views as of June 2016. This debate introduced an essential addition to the socioecological concept of resilience to ensure the concept's adequacy in the context of planning for social ends: that the resilience dynamic not be limited to a need to "bounce back" from a disaster event to a pre-disaster state, but to actually "bounce forward" toward a more advanced state of social justice than existed previously.

The value and sense of approaching resilience as a normative process, rather than a predetermined state of "future-proofing," are contested by many (Cox and Perry 2011). Elmqvist and his colleagues at the Stockholm Resilience Centre differentiate the concept of sustainability from resilience in terms of the normative dimension of the former, and not the latter. That is, whereas the concept of sustainable development articulates a goal toward which urban planning and development efforts should be oriented, resilience offers "a framing device for thinking about socioecological and urban systems … a new vocabulary for thinking about place-making based on evolutionary change rather than a linear pathway towards a single end-state" (Scott 2013: 430).[1] If Elmqvist and Scott are correct, and resilience thinking lacks normative content, then we can expect urban planning based on resilience thinking to limit the space we have to consider alternative practices, patterns, and horizons. If it is a concept without a normative thrust, resilience cannot allow us a process to delve deeply into the rewards obtained by groups and individuals for taking particular kinds of risks, and the different nature of the rewards they receive from safety. Without a normative edge, all it may offer is a way to develop and implement new strategies of insurance against rising risk. It may combat the risk society, but not necessarily in the same direction as urbanites hope, toward greater willingness to tolerate risk when they have a grasp of the social value of taking it.

Fainstein (2015) worries that, with this lack of a normative thrust, employing resilience "leaves the analyst with enormous mapping jobs and model-building challenges but provides little in the way of decision rules." Lacking a political, social, material grounding, resilience devolves into an empty signifier: "developing urban resilience as fashion or an empty concept without any real substance or potential to contribute to planning theory and practice" (Scott 2013: 431). By default, it is all too likely for resilience planning to side with the preservation of short-term safety and security, and forfeit opportunities to work within the essential urban urge for risk as well, in order to recognize and build on a fuller set of community strengths (Gibbs et al. 2015). Putting constraints on our thinking about what we can and cannot change about the way we organize ourselves in our cities is the biggest handicap of this application of resilience. To reorient urban resilience pragmatically, resilience needs to be situated within a political process in a community context.

## The scene of struggle

The pragmatic scene of struggle is the public domain in which dynamic arguments are offered and counter-offered, inquiries into the common social good are pursued and verbally attacked in public, and new tests and compromises developed. A widespread intellectual and popular dissatisfaction with the very notion of the "guaranteed city," along with the policies and practices in support of it, are strong testament to the essential social character of its opposite, the scene of struggle, a place in the city imbued with risk, and inviting to all at the same time. Thévenot refers to the state of "inquiétude", translatable as a state of worry, anxiety, or riskiness, as a state with particular fecundity for actors in the public realm.

The pluralism within our cities demands attention to difference as a precursor to effective engagement of any sort. In keeping with the central importance of authentic individuality to many urbanites today, the significance of difference only increases. By giving due recognition to this state of difference, and to the commensurate role to be played by a physical scene for a struggle of differences to occur, pragmatic method sets up the conditions for engagement with pluralism. This is not ideal, but it is as good as we can get. Within this scene of struggle, the core work to be accomplished is to recognize those distinctions that make a difference, that are relevant to the material stakes of the dispute at hand. This work is aided by recognizing amidst the conflicting plurality a relatively limited number of regimes of engagement, or bases upon which people are making claims for the worth or particular social value for their particular perspective. By recognizing the poles of gravity within the sea of social difference, it becomes possible for social actors to channel their energies into a small set of common groundings to define a new community, and act in that way in the common interest. As Thévenot (2014: 14) explains:

> A strong accent is nowadays put on differences. Yet the infinity of possible differences is channeled by the diverse grammars, each of them selecting a relevant kind of difference and way of differing. Because of this channeling, each construction offers a distinct mode of integrating differences – composing difference in the old sense of settling a disagreement that results in the composition of a plural common ground for the community.

This sense of hidden sites of commonality, and rationales to be discovered for common action in particular contexts, provides clear value and exciting prospects, inviting people to the scene of struggle, despite a certain inquiétude. Plurality is a recipe for conflict, certainly; and clashes can be intense. But seeking a sense of purpose and commonality within the boundaries of particular conditions eases actors into the possibility of integration and common interest: "Although clashes can be intense, common features shared by all orders of worth allow the possibility of local and temporal integration of differences between several orders of worth and result in what we analyzed as 'compromise'" (Thévenot 2014: 14; Boltanski and Thévenot 2006 [1991]). The pragmatic approach takes the isolation and irreducible difference out of pluralist individualism: "plurality does not mean that we are limited to being separate individuals with irreducible subjective interests. Rather it means that we seek to discover some common ground to reconcile differences through debate, conversation and dialogue" (Bernstein 1983: 223). This, in effect, is the meaning of the public realm defined by Hannah Arendt (1998 [1958]: 53) in this way:

> The public realm, as the common world, gathers us together and yet prevents our falling over each other, so to speak. What makes mass society so difficult to bear is not the number of people involved, or at least not primarily, but the fact that the world between them has lost its power to gather them together, to relate and to separate them. The weirdness of this situation resembles a spiritualistic séance where a number of people gathered around a table might suddenly, through some magic trick, see the table vanish from their midst, so that two persons sitting opposite each other were no longer separated but also would be entirely unrelated to each other by anything tangible.

Without calling it by the term, Dewey recognized the scene of struggle in the public realm as key to locating, and coming to understand and support political life in a democracy, holding that "the aim of political life is to make it possible for citizens to identify their shared interests and find effective mechanisms for pursuing them" (MacGilvray 2010: 33). Dewey also supported a diversity of possible allegiances and (without referring to them as such) orders of worth as a route to individual growth:

> we should promote forms of social organization that allow for the greatest range and variety of voluntary associations, on the grounds that this kind of pluralism will leave room for the cultivation of the greatest range and variety of social goods.
>
> (ibid.: 39)

Dewey (1927: 147) went further in *The Public and Its Problems* to support the interplay of different organizations, along similar lines:

> From the standpoint of the groups, [the democratic idea] demands liberation of the potentialities of members of a group in harmony with the interests and goods which are common. Since every individual is a member of many groups, this specification cannot be fulfilled except when different groups interact flexibly and fully in connection with other groups.

This kind of interaction entails a commitment to democracy. Within the intricate inner workings of this commitment to democracy, people act by juggling different modes of determining worth and proper action. Critical pragmatists take a close look at these risky interactions and, by dissecting them, aim to better channel their results toward action in the common interest.

Ash Amin (2008: 18–19) refers to the urban attraction to such risk as the positive experience of "conviviality," which he describes as "an everyday virtue of living with difference based on the direct experience of multi-culture, getting around the mainstream instinct to deny minorities the right to be different or to require sameness or conformity from them." Urbane actors, when they abide by the social contract, respect the rights of others to peacefully coexist. This is a crucial skill in the city, where citizens do not always have the ability to separate themselves physically; there is a need to find a social means to security. While some argue that cities offer fewer opportunities for serious efforts at multiculturalism than would be required to overcome deep-seated fears of the other, Amin suggests the momentary "brush with multiplicity" in the city's public spaces, an "aesthetic disruption ... in the spirit of reinventing the ties that bind" (ibid.: 16). The promise of realizing greater value from multiculturalism in the future may be enough to maintain the ideal in the urban body politic.

Pattaroni and Pedrazzini (2010) join others in recognizing that a growing cadre of cities is closing the gap between a perceived need to ensure security and a lack of trust in a socially and civically guaranteed sense of civil security. This gap is closing in the wrong direction, so far as the democratic axis is concerned: through the emergence of a private, partially digital, security industry. Beyond the security industry per se, many retreat to the supposed safety of walls, both of the physical and figurative variety. Policy, planning, and management strategies show a preference for separating those who might have differences that would threaten or be perceived as threatening to others. Individuals, too, see the lack of a social contract and retreat to their own identity groups, for their own protection and sense of security. This represents a loss of faith in the promise of the commons. If we admit failure of the social contract, even at the scale of our cities and communities, is there anything left for the city to do but build walls? There can be no urban commons in this formulation.

We can see policy decisions related to risk and safety coming down on the side of enclosure as well as on the side of open-minded space, expanding and restricting the resilience of cities and the extent of their scene of struggle, in placemaking settings in cities everywhere. Take the example of community gardening, for example, a popular component of placemaking strategies for its ability to connect people with one another, and with the nonhuman nature of their home. In Figures 6.1 and 6.2, the contrast in decision-making is clear. In Figure 6.1, the Cottonwood Community Garden was established in 1991 with a view of providing gardening space and a green haven for all in its busy, urban East Vancouver neighbourhood. Figure 6.2 shows the French écoquartier at Plateau de Haye, a redevelopment of a large social housing project in Nancy, France. Set amidst a large nature park, on the hilltop above Nancy, the residents of the social housing apartments wanted space to garden. While the state developers initially designed a more open community garden, complaints of theft pushed them to build tall chainlink fences around each household's allotment.

At the neighbourhood scale, aside from the most obvious examples of formally "gated communities," it is at the borders of new placemade neighbourhoods that we can most obviously see the evidence of decisions being made either on the side of the commons or on the side of enclosure. Are there open-minded spaces being created at the borders of these new districts? Or, instead, are these districts being created as entirely inward-looking, private enclaves, with little attention to or sense of responsibility for the zone of interaction with the connecting communities? A range of outcomes can be found. A few examples are shown in Figures 6.3–6.6. In Figure 6.3, again from Plateau de Haye, the starting point for the écoquartier is a modernist housing project at the end of its life. The housing project is known as "La Barre" due to its shape and size, 1.3 km long. As part of the redevelopment, the middle chunk of the building was demolished and made into open space. Residents of this portion of the building were moved into new apartments and townhomes built within a greenspace-dominated landscape that is now outfitted with trails and clearings to make it more inviting for public use. The remaining two smaller buildings had interior renovations. The result is a district with many more edges facing nature and a greater focus on exterior semi-private as well as public, natural spaces.

In Figure 6.4, the redevelopment of the Stockholm Royal Seaport is shown with one of its prominent neighbours, the older gasworks structures that once productively occupied the whole site. These structures literally loom over the new ecodistrict and are closed off to residents and visitors. On the other side, the site is flanked by a large nature preserve, owned by the king. Both these

*Figure 6.1* Cottonwood Community Garden, East Vancouver. Community gardening
has become a popular action to take for mutual sustainability and justice
goals in cities in many parts of the world. This is Cottonwood Community
Garden, the oldest in the City of Vancouver, established in 1991. Across the
street, at the rear of the frame, is the Greater Vancouver Food Bank.
Source: Photo, Blair Bellerose.

factors create a sense of isolation from the city beyond. Figure 6.5 demonstrates
the active work and investment of the public development company of Hamburg, Germany, to invite mixing at one of the key entry points to the Hafen-
City district, a new sustainable district intended to become the city's "next
downtown." A new metro station has been added, with free access to transit
for an initial period as an additional draw. The metro station is adjacent to the
new HafenCity University. Public bike share racks are available, and other
seasonal and ongoing attractions are regularly promoted. As a final example of
the diversity of practices at the zone of either connecting or erecting barriers
between the new ecocity and the existing city, Figure 6.6 shows part of the
Västra Hamnen district of Malmö, Sweden. At this large site whose phased
construction began in 2001, this large park and children's play area, interior to
the site, serve as an important mixing zone. With play equipment and gardens

*Figure 6.2* Allotment gardens, Plateau de Haye housing project, Nancy, France. Among many other justifications, the popularity of community gardening reflects an order of worth oriented around valuing a material city with productive life-sustaining potential, if properly tended, and valuing a good life within the limitations and provisions of the garden. Here, an ethic of enclosure within a space intended for community gardening. The fences and locks were put on individual allotment gardens following accusations of theft from gardens.
Source: Photo, the author.

designed to be suitable for different age groups, from the very young to older adults, this space balances open-mindedness with designed-in purpose. In addition to the waterfront promenade, which the district opens up to Malmö residents, and which has proved very popular, this park provides additional reason to linger and mix. (I would also hasten to add that the risk level of the playground equipment itself, in terms of personal safety of those daring to play on it, is rather higher than what I have experienced in other cities. No wonder the children love it.)

The operating principle in the task to securitize the city is one of autonomous defensiveness of each person against the other, and each part of the city against every other. Each reinforces the sense of separation from the other. The

*Figure 6.3* Plateau de Haye, Nancy, France. In this hilltop district above the city of Nancy, a modernist social housing project is being transformed into an éco-quartier with federal subsidies and support. The original social housing structure, referred to as "La Barre," originally stretched 1.3 km. The redevelopment project retains the building, with extensive interior renovations, and a swath demolished in the middle. Extensive gardens and natural areas are part of the redesign, as can be seen here, as well as new stacked townhouses with wood exteriors. The other neighbour on this site is a federal prison, which may add to the pervasive use of fences in this district.
Source: Photo, the author.

common has no identifiable value, only risk. Walls, exclusionary zoning, barricades, fences, and security guards play the separating functions in the city that, in a rural society of fear, is played by the distance between warring villages. Securitizing the city feeds a politics of insecurity. The sense of individuals' impotence to play any role in their own security is eroded by the experience of securitization practices. It brings to mind the notion that threats are pervasive whenever these people and structures are seen, reinforced by the media which reports only on crime and threats of crime. This represents a profound reshaping of the social realities of city life. Insecurity leads to segregation, and the cycle repeats.

To work in the service of growing the commons in the city, urbanites need to abandon the notion of the guaranteed city as a feasible, reasonable policy objective. They may find value in turning more toward the more open-minded concept of resilience, which permits a pursuit of social learning in the interests of adaptive capacity, and does not turn away entirely from dispute and risk. We depend on our social relationships to a large extent for our sense of self. When we are unable to feel any sense of solidarity within them, or when our membership is revoked through some turn of events like ill health or unemployment, or the end of a personal relationship, we may struggle to recognize our own selves. The

*Figure 6.4* (a) Stockholm Royal Seaport, Sweden; (b) the former gasworks infrastructure
looms large over this large redevelopment project in Stockholm.
Source: Photo, Charling Li.

struggle for recognition in the face of such disempowerment, absence of
capacity, absence of decency, or freedom from humiliation beyond justice
strictly defined, that is the resulting social urge. This struggle can often take the
form of anger, denunciation of others, and too often, violence.

Harking back to our thought experiment with the street protest for sustainability and justice in Chapter 1, people retreat to defining themselves in terms
of which side of the fence they are on, and define for their own sense of safety

*Figure 6.5* Edge of HafenCity, Hamburg, Germany. In HafenCity redevelopment project, attempts to create open and lively spaces at the borders with the existing city include a new underground metro station, adjacent to the new public HafenCity University, with access to the public bike share system as well as other amusements.

Source: Photo, Charling Li.

*Figure 6.6* Västra Hamnen playground mixing ground, Malmö, Sweden. In Västra Hamnen, the redeveloped neighbourhood opens up the harbour to the city, providing an attractive place for recreation. In the centre of the neighbourhood sits another open space, with expansive play areas oriented toward different age groups, from very young to adults.

Source: Photo, Charling Li.

and justice which is the good side. The potential and creativity within the diversity of perspectives, justices, sustainability visions, and action strategies among the agitants on the downslope side of the fence, are lost on the participants, in this scenario of the securitized city. Because it is a simpler, non-knowledge way of understanding safety, it remains as a dangerous fall-back position at all times. Even after one has decided to engage in politics, it is always more challenging, more risky, and more demanding of personality, to learn to engage with those who see the struggle differently than you do.

## Towards a greater commoning of the city

The existence of a commons implies the existence of a community, and this community can only operate effectively according to a common set of practices and common knowledge. Studies of commoning, therefore, investigate how the commons are created and maintained by these actions, knowledges, and relationships, over time, across space, and through conflicts and struggles (Chatterton 2010; Fournier 2013). The work of commoning is work in direct opposition to that of enclosure (Linebaugh 2008), or privatization, which is an act of the capitalist urban development process, seeking to "forcibly separate people from whatever access to social wealth they have which is not mediated by competitive markets and money as capital" (de Angelis 2007: 144). As the dynamics of neoliberalism push this act of enclosure into new domains, beyond land and natural resources to intangible intellectual resources, genetic material, and electronic transmission space, the work at hand for commoning expands, too. Commoning work thus becomes a dynamic set of political activities which expand options for urban life in creative new ways, much more than simply reflecting older ways of managing limited resources communally. Jeffrey et al. (2012: 1263) position the work of commoning as "enclosure's other" and view it as essential to advancing community work of all kinds: "To the extent the re-thinking of community has installed itself at the heart of international philosophical debates, the idea of the 'common' or 'commons' has revitalised attempts to examine the nature of collective political projects."

For Harvey (2012), to theorize the material culture of a given urban commons it is necessary to raise symbolic questions about who are the uncommon against which the common resources are named and imagined. There must be some who do not fit what seems to be normal. Harvey is working backwards here, from a realized commons to the irreducible differences that may have come together to create the commons. Like other justice theorists and activists, he is impatient for the work that comes after the creation of the commons. Referring specifically to the work needed in American cities, one activist puts it this way:

> Imagine dialogues on neighborhood development and urban design occurring among protest participants. Imagine planned public talks hosted on neighbors' stoops or in the foyers of housing projects. Imagine democratized approaches to urban planning that begin with the people, not the

corporate class. Imagine the embedding of urban planners within movement collectives combatting anti-black racism and state-sanctioned violence from Ferguson to Flatbush. That type of work is characteristic of the critical first steps needed to inform the creation of the "just" city.

(Moore 2015: 20)

In other words, wherever a common place is defined in a city (a street, a park, a school, a clinic), there is a right that is put into place in the form of a law, a regulation, or a rule backed by either civil or penal sanctions. This provides a focus for the action of the community formed to defend and protect that right. This is an administrative form of citizenship. Wherever there can be an interpretive contradiction in which people stand up to argue for their right to the city and against how they are being represented is unjust – this is an act of citizenship (Isin 2012). This work is the work of commoning and sits solidly within a context of individual authenticity, building authentic communities without losing the uniqueness of the individuals who constitute them.

Within each distinct regime of engagement in the pragmatic view, communication takes shape in an attempt to make personal concerns common within a group. This is required "to alleviate the tension of living both together and in person" (Blokker and Brighenti 2011: 8). Each involves the testing of one's belief against one's view of authenticity, and the composition of a plurality of voices and constructions of the common good, common places, common sites of meaning. And each different regime of engagement has its specific advantages and disadvantages where community building objectives are concerned. The distinction among these regimes of engagement reveals the shortcomings of modern constructions of individuals, debates, and publics, including public spaces. Debates are differentially invested with different sets of analytical tools. If we hope to keep public debate within the realm of the constructive, let alone socially transformative, and avoid settling into the regressive and reactionary, we need to maintain our awareness of the different forms of engagement.

Not all urban citizens are prepared to generalize, to think of ourselves as capable self-maximizing individuals (even in some idealized way), and keep the prospective Great Community in our sights. Still, we all have the potential to be political actors. People disagree, both with one another and within ourselves. It is important to understand the modes and contexts of these disagreements in order to advance objectives of increased engagement in the public sphere, and possibly to arrive at transformative alternatives. Doing this

sheds light on the same core problem raised by living together: the problematic integration of a wide variety of modes of evaluating and coordinating one's own behavior, as well as coordinating with other actors. If one takes seriously the strong discrepancy between public commitments and more personal ones, the problem of how to integrate the full range of commitments comes into view.

(Thévenot 2014: 12)

In working toward this integration, we gain the tools to register the value of protecting and reinforcing human and social diversity within our ideals of the Great Community.

Thus a view across individuals and their authenticity, and across communities and their work to generate a sense of the commons, becomes necessary. In order to ensure that these common places can be identified, and open to view in communities, in individuals, and in public, ongoing engagement is necessary. This becomes more challenging, more important, and more rewarding, the more diversity is contained within the scene of struggle. If urban citizens become too diverse and too mobile, they lose civic memory, increasing the requirements for social learning. If cities are not diverse, open and mobile enough, they lose the convivial spirit of the public sphere as they lose their ability to maintain population and attract newcomers. If they take on too much local self-governance, they risk losing the sense of security, comfort, and coordination of central authority. If they have too little self-governance, they run into conflict over a lack of local resilience and low levels of direct and representative forms of democracy.

Importantly, the demand to translate justifications into publicly acceptable forms is not an act of selflessness in view of the common interest. It is a way to make things happen. A particular way of thinking and proposed course of action garners support and succeeds in implementation when, pragmatically speaking, it 'hitches itself' to an appealing, prevailing discourse of the common good. This argument over how best to characterize the common good is foaming with power, and sits entirely within baser principles of narrow interests and manipulation. Those who are best equipped to manipulate public opinion, "the incensed and the articulate" (Carson and Martin 1999: 57), will aim to privilege their narrow interests over others. Dryzek and Niemeyer (2010: 111) refer to this as "a cognitively cheap solution to the problem of constructing preferences in relation to complex problems."

At the same time, the attraction of manipulating the public sphere in order to favour narrow interests is not the only attraction. There are others. The rise of the 'sharing economy' and 'maker culture' suggests some of the directions being taken by those who are not satisfied with the strictly individualist, privatized appeal of mainstream urban culture. The idea is to revalue sharing skills and positional goods with others on a "community basis," in the case of the sharing or solidarity economy (Gardin 2006). In the case of maker culture, the idea is to revalue craft and task knowing, from an experiential, community-of-practice, as opposed to strictly market-based, perspective (Amin and Roberts 2008). Regardless of the extent to which these social trends do or do not meet their intended goals, they are significant in terms of the challenge they pose to the privileging of mainstream market values and the value they instead assert for reciprocity, sufficiency, and mastery of an art, craft, or topic.

The commons is a physical and a figurative space for the work of communities. This is real work, in dire need of greater recognition. To move toward a pragmatic ideal of a justice-seeking culture, public spaces need to be created and maintained as open-minded spaces of the commons, with attention paid to the habits that are created by the act of spending time in these spaces. This all suggests a need for greater autonomy within these spaces. Fair enough. At the same time, however, this work can be taken on in earnest by many more urbanites without forcing a relinquishment of all the pleasures and securities that the market order of worth can provide. As Fournier (2013: 451) recognizes:

> We cannot walk out of the market without access to other resources. We cannot, for example, stop buying food without having access to land on which to grow our own, knowledge of how to do so or networks through which we could exchange that food for whatever skills, products, knowledge or help we can offer. To escape dependency on commodity markets, we need to reconstitute resources, relations and knowledge.

Community engagement, community use, production, and reproduction can fill a void in the understanding of daily life within the dominant market rationality that can help urbanites accommodate what, by market logic alone, might be unacceptable risk.

The concept of resilience may be of help in moving forward into this risky terrain. Growing interest in partnerships and intermediary objects may also be part of the same solution. Partnership approaches to changing the organization and management of space, leveraging the capacity and position of local government and private sector with that of community actors, are often part of more just solutions. But not always. In particular, while partnership approaches are justified using a language of efficiency and the distribution of risk to the private sector, as partnerships have borne out in urban redevelopment practice, "when 'efficiencies' fail there is a tendency to fall back on the public purse" (Krawchenko and Stoney 2011: 77). Intermediary objects may be organizations or vehicles for communication that function within or outside of a formal partnership. These agents reduce unacceptable risk to those engaging in the commons by translating unaccustomed ideas, attachments, and feelings into a recognizable tongue. Sometimes the common place itself accomplishes this translation of the very personal to the common, due to the way the setting itself, or a particular building, tree, river, park, or other object in the urban environment compels the attachment of many. Other times, an intermediary able to employ *koinos topos* (common place), in terms of a general theme associated with a common place, can bridge the divide from the specific out to the commons. In both cases, "communication requires a transformation to express familiar concerns and attachments through common-places" (Thévenot 2014: 20).

## Conclusion

Life in the city, both in its mainstream neoliberal mode and in community work at the neighbourhood scale "has its own walls" and exclusionary practices (Jeffrey et al. 2012: 1254). The scene of struggle in the very places and times at which the commons is constructed and preserved, is itself not immune to particular forms and politics of enclosure, separation and division. As we saw in examples of community gardens and urban sustainable redevelopment projects, an attempt to generate the new commons may in itself provoke new feelings and forms of resistance and enclosure, whether by design or by practice. The forces of enclosure and privatized safety become stronger in proportion to how weak our sense of the commons becomes. As we lose this sense, in spite of its risk, we also lose the ability to reasonably justify the pursuit of social and sustainability justice. Conceptualizing the important role of risk in our understanding of what is valuable about cities, and what is valuable about urban ways of life, may support efforts to struggle against the securitization of "the guaranteed city." The notion of urban resilience, as a process and set of tools that can be loaded with the social and normative thrust of progress through shock, struggle, and stress, may provide needed fuel to the struggle to maintain what is crucially risky about urban life. The contradiction between risk and security, and risk and reward, represents an important tension in urban culture and politics, and one that may wind up supporting the emergence of a new, urban, model of justice.

Thinking about the evolving urban mode of life, in Chapters 4–6 of this book we have traced possibilities for new kinds of competing claims vying for a sense of what counts for justice in the contemporary urban public realm. We have examined the way in which urbanity offers a new understanding of the common good that depends upon a multiple, shifting, ever-uprooted sense of authenticity. And we have argued that this sense of achieving an authentic life in the city may reopen older debates about community and the common good. We have looked at the shifting, urbanizing terrain of citizen engagement and empowerment and seen that the qualities of and arguments for public and community engagement are open to more pragmatic, urban reinterpretation. Tracking engagement and empowerment in communities is an important dimension of dynamic action and argumentation in the public realm. Finally, we have examined the way in which the struggle to reproduce "the guaranteed city" directly opposes the pragmatic struggle to maintain what is crucially risky about urban life, by generating and maintaining the commons. Taken together, we have built a case for the consideration of a new, uniquely urban regime of engagement for representing justice in the public sphere that revolves around: authenticity and engagement, empowerment and community, and risk and resilience. Justice in the contemporary city is available to those who still take to the commons of the public sphere to make sense, act, and form beliefs in the maelstrom of isolation, disempowerment, privatization, and risk that constitutes so much of modern urban life. In the concluding chapter, we review this case.

## Note

1 The insinuation that "sustainability" compels planning action toward a single pre-determined end state is highly suspect. For a number of these suspicions, see Chapter 2.

# References

Adger, W. Neil. "Social Aspects of Adaptive Capacity." In *Climate Change, Adaptive Capacity and Development*, edited by Joel Smith, Richard Klein, and Saleemul Huq. London: Imperial College Press, 2003, pp. 29–49.

Adorno, Theodor. *Essays on Music*. Berkeley, CA: University of California Press, 2002 [1941].

Amin, Ash. "Collective Culture and Urban Public Space." *City* 12(1) (2008): 5–24.

Amin, Ash and Joanne Roberts. "Knowing in Action: Beyond Communities of Practice." *Research Policy* 37(2) (2008): 353–369.

Arendt, Hannah. *The Human Condition*. Chicago: University of Chicago Press, 1998 [1958].

Beck, Ulrich. *Risk Society: Towards a New Modernity*. London: Sage, 1992.

Beck, Ulrich. *World at Risk*. Cambridge: Polity, 2009.

Bernstein, Richard J. *Beyond Objectivism and Relativism: Science, Hermeneutics, and Praxis*. Philadelphia, PA: University of Pennsylvania Press, 1983.

Blokker, Paul and Andrea Brighenti. "An Interview with Laurent Thévenot: On Engagement, Critique, Commonality, and Power." *European Journal of Social Theory* 14(3) (2011): 383–400.

Boltanski, Luc and Laurent Thévenot. *On Justification: Economies of Worth*. Princeton, NJ: Princeton University Press, 2006 [1991].

Breviglieri, Marc. "Une brèche critique dans la 'ville garantie'? Espaces intercalaires et architectures d'usage." In *De la différence urbaine. Le quartier des Grottes/Genève*, edited by Elena Cogato Lanza, Luca Pattaroni, Mischa Piraud, and Barbara Tirone. Geneva: MētisPresses, 2013, pp. 213–236.

Brugmann, Jeb. *Welcome to the Urban Revolution: How Cities Are Changing the World*. New York: Bloomsbury Press, 2009.

Carson, Lyn and Brian Martin. *Random Selection in Politics*. Westport, CT: Praeger, 1999.

Chatterton, Paul. "Seeking the Urban Common." *City* 14(6) (2010): 625–628.

Cox, Robin S. and Karen-Marie E. Perry. "Like a Fish out of Water: Reconsidering Disaster Recovery and the Role of Place and Social Capital in Community Disaster Resilience." *American Journal of Community Psychology* 48 (2011): 395–411.

Davoudi, Simin, Keith Shaw, L. Jamilla Haider, *et al.* "Resilience: A Bridging Concept or a Dead End?" *Planning Theory & Practice*, 13(2) (2012): 299–333.

De Angelis, Massimo. *The Beginning of History: Value Struggles and Global Capital*. London: Pluto Press, 2007.

Dewey, John. *The Public and Its Problems*. New York: Henry Holt & Co., 1927.

Dryzek, John and Simon Niemeyer. *Foundations and Frontiers of Deliberative Governance*. Oxford: Oxford University Press, 2010.

Fainstein, Susan. "Resilience and Justice." *International Journal of Urban and Regional Research* 39(1) (2015): 157–167.

Florida, Richard. *The Rise of the Creative Class ... and How It's Transforming Work, Leisure, Community and Everyday Life*. New York: Basic Books, 2002.

Foucault, Michel. *Governmentality*, trans. Rosi Braidotti, revised by Colin Gordon. In G. Burchell, C. Gordon and P. Miller (eds.) *The Foucault Effect: Studies in Governmentality*. Chicago: University of Chicago Press, 1991, pp. 87–104.

Fournier, Valerie. "Commoning: On the Social Organisation of the Commons." *Management* 16(4) (2013): 433–453.

Friedmann, John. *The Prospect of Cities*. Minneapolis, MN: University of Minnesota Press, 2002.

Gardin, Laurent. *Les Initiatives solidaires: La réciprocité face au marché et à l'état.* Paris: Erès, 2006.

Gibbs, Lisa, Lousie Harms, Sarah Howell-Meurs, *et al.* "Community Well-Being: Applications for a Disaster Context." *Australian Journal of Emergency Management* 30(3) (2015): 20–24.

Gleeson, Brendan. *The Urban Condition.* London: Routledge, 2015.

Gorz, Andre. *The Immaterial.* Chicago: University of Chicago Press, 2010.

Hamann, Ralph and Kurt April. "On the Role and Capabilities of Collaborative Intermediary Organisations in Urban Sustainability Transitions." *Journal of Cleaner Production* 50 (2013): 12–21.

Harvey, David. *Rebel Cities.* London: Verso, 2012.

Isin, Engin. "Citizens Without Nations." *Environment and Planning D: Society and Space* 30(1) (2012): 450–467.

Jeffrey, Alex, Colin McFarlane, and Alex Vasudevan. "Rethinking Enclosure: Space, Subjectivity and the Commons." *Antipode* 44(4) (2012): 1247–1267.

Krawchenko, Tamara and Christopher Stoney. "Public Private Partnerships and the Public Interest: A Case Study of Ottawa's Lansdowne Park Development." *ANSERJ* 2(2) (2011): 74–90.

Linebaugh, Peter. *The Magna Carta Manifesto: Liberties and Commons for All.* London: Verso, 2008.

MacGilvray, Eric. "Dewey's Public." *Contemporary Pragmatism* 7(1) (2010): 31–47.

Merrifield, Andy. *The New Urban Question.* London: Pluto Press, 2014.

Moore, Darnell. "Urban Spaces and the Mattering of Black Lives." In *The Just City Essays: 26 Visions for Urban Equity, Inclusion and Opportunity.* Edited by Toni Griffin, Ariella Cohen, and David Maddox. V.1. New York: J. Max Bond Center on Design, Spitzer School of Architecture, City College of New York, Next City and The Nature of Cities, 2015, pp. 18–20.

Norris, Fran H., Susan P. Stevens, Betty Pfefferbaum, Karen F. Wyche, and Rose L. Pfefferbaum "Community Resilience as a Metaphor, Theory, Set of Capacities, and Strategy for Disaster Readiness." *American Journal of Community Psychology* 41 (2008): 127–150.

Pattaroni, Luca and Yves Pedrazzini."Insecurity and Segregation: Rejecting an Urbanism of Fear." In *Cities: Steering Towards Sustainability*, edited by Pierre Jacquet, Rajendra Pauchari, and Laurence Tubiana. Delhi: TERI Press, 2010, pp. 177–187.

Puri, Belle. "Walking Tours Exploit Vancouver's Downtown Eastside, Advocates Say." *Vancouver Sun.* 2016. Available at: www.cbc.ca/news/canada/british-columbia/wa lking-tours-exploit-vancouver-s-downtown-eastside-advocates-say-1.3713724 (accessed 18 August 2016).

Robinson, Jennifer. *Ordinary Cities.* London: Routledge, 2006.

Ross, Andrew. *Bird on Fire: Lessons from the World's Least Sustainable City.* New York: Oxford University Press, 2011.

Scott, Mark. "Editorial." *Planning Theory and Practice* 14(4) (2013): 429–432.

Sennett, Richard. *The Uses of Disorder.* New York: Vintage, 1970.

Sennett, Richard. *The Fall of Public Man.* New York: Random House, 1978.

Thévenot, Laurent. "Voicing Concern and Difference: From Public Space to Common Places." *European Journal of Cultural and Political Sociology* 1(1) (2014): 7–34.

Watson, Sophie. *City Publics.* New York: Routledge, 2006.

Zolli, Andrew. "Learning to Bounce Back." *New York Times.* opinion pages, 2012 (Nov. 2). Available at: www.nytimes.com/2012/11/03/opinion/forget-sustainabili ty-its-about-resilience.html?pagewanted=all&_r=1 (accessed 18 August 2016).

# 7 Conclusion

## A better urban life to be lived

In this book, we have sought a closer look at the workings of individuals and communities within the contemporary city, to see what purposes our current values of authenticity and empowerment, community and engagement, and risk and resilience might serve. We have restated the urban question at this moment of crises of societal injustice and environmental unsustainability in pragmatic terms. Namely, what gives in today's resurgence of urban habitats and urbanist values when it comes to the prospect of pursuing sustainability and justice more authentically, more democratically, and with more daring?

The goal of urban development appears around the world in a singular, growth-oriented fashion – like a single neon sign that blinks OPEN, bright enough to be seen from space. It is not difficult to envision the urban utopia presently on offer. Castells (1977) was right: the ideal city, judged by the aggregate of decisions being made, is an ever-expanding opportunity structure for consumption and accumulation at both grand and tiny scales. In this image, cities can be determined to have earned economic success based on measures of recovery and growth, even if no new jobs are created. Flexible, temporary employment is the only kind of job security on offer. It is difficult for most individuals to consider their voice as important in this version of a bigger urban picture (Standing 2011). And Banfield (1968) and Zukin (1995) were right, too: the ideal city is eternally spreading out over the landscape, and it is also constantly growing up in space and class terms, revitalizing and rebuilding urban spaces to let more capital in. A sense of individual disempowerment pervades, and along with it, a cynicism about any prospects for transforming society in a direction other than further economic growth, further segregation of interests and relegation of values to private domains. In our full-world context, where the prospects for transformation seem to be increasingly limited by the demands of capital, controlled by a precious few, why bother?

With this particular urban ideal being advanced so persistently and pervasively, critique seems in much greater demand than alternate utopias. Critique brings the faults and failures of this global urban ideal into the light where they can be challenged. Earnest attempts to construct another fully realized kind of urban utopia risk either being laughed at, or else co-opted into the dominant ideal. To wit, cities' sustainability plans assume and celebrate constant growth in mobile capital. Placemade urban design emphasizes inclusive amenities while ignoring the displacement and exclusion of former inhabitants. Urban citizens rally for greater housing affordability, yet cling to personal retirement dreams based on property speculation. The neon sign blinks on.

There is no denying that neoliberal forces such as capitalism, privatization, globalization, individualism, growth, and securitization are relentlessly at work in contemporary cities. The urban celebration rides this wave even as it turns its back on the ways in which these forces dismantle and make a mockery of many higher aspirations for the city. And yet, neoliberalism may well be a many-headed monster, but it is no medusa that has turned all other creative forces to stone.

## A pragmatic approach to urban potential

In the pragmatic approach taken in this book, we have followed the critique of totalizing claims made within planning theory by Watson (2009), Parnell and Robinson (2012) and Hoch (2007). In other words, despite the powerful thrust of the critique of neoliberalism, many spheres of activity in our daily lives are not colonized by the market (Gibson-Graham 2006). We have seen a number of different trends "boiling over" in the experience of the city, to borrow a metaphor from William James. Capitalism depends, now as ever, upon a community "fix" (De Angelis 2013). Taking an interest in the dynamics of community, as this fix sets in, provides an opportunity to redirect the current of social practice toward more pragmatic and potentially more progressive ends. Power is not only exercised upon people; people are always exercising different forms of power upon one another and upon the material conditions of their life. The common good, the community, and the roles and places that we may authentically hold within it are made meaningful by pragmatic negotiation and compromise-seeking, much more than they are made meaningful by utopian conjectures. Seeking compromises, common purpose, and commonality more generally is possible against the backdrop of particular material context, as well as within a particular mixture of perspectives and models of worth. New compromises are always available, and new perspectives can always be forged, so long as we retain habits and abilities of acting in community, with one another, toward better, more workable visions of our public interests.

This period of resurgence of city thinking, city living, and city leadership is exciting for urbanists. The contemporary urban celebration, dissected in Chapter 3, shows us that something momentous is afoot in our understanding of our place in the city. The paucity of social critique of the urban celebration shows us that this understanding has yet to be adequately articulated. In this book, I do not seek to join the celebratory chorus. On the other hand, I do not seek to denounce this newfound interest in urbanism, either; this passion is a part of me, too. I look to the urban celebration, in urban scholarship and in urban observation and experience, as well as to the critique that exists, for more understanding of the possibility for sustainability and justice.

A pragmatic approach guides me in providing a sense of reasoned hope in the present moment for progress into a future version of utopia that is not entirely far-fetched. It offers the tools that a critical and engaged community needs in order to move closer toward a city of sustainability and justice: a sense of the value of multiple authenticities, of empowerment within a community of inquirers, and of resilience within a persistent scenario of risk. A pragmatic approach gives the analyst the ability to dissect the justifications being offered. Pragmatism as a method of approaching justice and sustainability means, in large part, promoting compromises which advance justice and sustainability over other ideals such as individual freedom, procedural participation, efficiency, and security. We strike compromises based on our commitment to these and other conceptions of significant urban values on a daily basis. The fact that some values rise higher than others, on a more predictable basis, on the list of political and popular priorities has to do with the quality of our communities, the quality of our discourse, and the quality of our public spaces, as much as it has to do with the extrinsic demands of capital. The pragmatic wager, as per Dewey's notion of the Great Community, is that in improving our communities, elevating our discourse, and (now drawing more upon the critical pragmatists than upon Dewey himself) improving our public sphere, the citizenry has the means to justify and advance the just and sustainable city.

The next question is whether the contemporary city is, in fact, helping or hindering this ongoing, pragmatic, democratic effort. Critical urbanists have expressed grave doubts. When considering the content of the contemporary urban celebration, critical urbanists conclude that we are celebrating the city for all the wrong reasons. We took up this startling realization in Chapter 3. Within critical social science, we parsed out three key traps that the celebration of the city has fallen into: the local trap, the empowerment trap, and the community trap. Each articulation of the injustice that inheres within the thrust of what is currently celebrated about the city raises grave concerns for the future sustainability of the city. Even as we revalue local-scaled social action, and reclaim urban neighbourhoods through work like community gardening, we may be constructing new walls that prevent more powerful and more just political coalitions that jump this local scale. Even as we speak out against the concentration of capital and power at public meetings, create neighbourhood campaigns to collect signatures and energies to make our views heard in social media, and engage in participatory politics of different varieties, we may be victim to a new form of governmentality that keeps us in sanctioned partici-patory channels where our input can be contained within the range of what is safe to existing power structures. Even as we pursue voluntary, community work, in the interests of contributing to a vibrant street life, we may find our sharing economy to be decreasing the value of sharing, and opening up new arenas into which capital can monetize more aspects of urban life. Even as we aim to pursue justice and sustainability, we are headed off at each juncture.

These critiques and warnings merit serious pause. Pragmatically, however, they should not condemn our work to recreate the city in conditions more favourable to sustainability and justice. The fact and tendencies of injustice and unsustainability in the city are reason enough to suspend celebrations, but not reason enough to abandon the urban ship. In Chapters 4–6, pragmatic reasoning and strategies, and reinforcing evidence and examples, were offered. We sought the possibility that a new understanding of the city may offer justice and sustainability through ordinary, local-scale action, through citizen empowerment, through community work, and through an open engagement with risk and resilience.

The task of constructing a pragmatic view of urban justice and sustainability entails adopting an analytical stance that places great value on the engagement of diverse individuals in the civic sphere, and on understanding the specific material as well as the social and political conditions in which this engagement takes place. Rather than beginning with an understanding of power being exercised from above and resisted from below, a pragmatic approach seeks to unveil the power being employed by all actors, in context, and how certain regimes of engagement and strategies of justification gain power over time. Ultimately, the question explored in Chapters 4–6 was whether there is cause to identify the current urban moment as offering a new regime of engagement, a new model of specifically urban justice, that could be elevated to guide urbanism in the direction of justice and sustainability. Within this prospective urban regime of engagement, Chapter 4 identified the role of urban authenticity, valuing identity, personality, and the ability to make life plans as well as adjustments to these plans without losing one's sense of self. Chapter 5 examined the urban work of empowerment, and new forms of social learning and community-building in the public sphere. Chapter 6 looked at the role played by risk in cities as a third possibly unique urban value to be upheld.

In each case, the argument was exploratory, and cognizant of the pitfalls and "traps" that abound to confound justice and sustainability-seeking in the city. At the same time, the contemporary urban moment is tied historically to reasons that have already been articulated, to value these very concepts. This history includes values such as human "rootedness," holding the thread of utopian visions as we pursue concrete projects in public space, seeing the social community as a space of learning and constructing the essential pre-conditions of democracy, and social resilience as a means to rise above the demand for security of person and property in a context of risk. With a pragmatic transformation toward the specific potential offered for each of these values in the contemporary urban condition, we may find a path to urban sustainability and justice.

Serving our community ideals need not be premised upon ironing out differences, as implied by the notion of the Great Society, nor upon the physical determinism of certain approaches to placemaking, nor upon following procedural notions of participation and empowerment. Better results for arguments in the service of justice and sustainability, instead, demand recognition of the many ways in which people might engage in the public sphere. Meeting others within the public sphere serves diverse individual and social needs, in a range of forms and orientations of membership and attachment. As a set of complex social processes, habits, and beliefs, community-building requires a commitment to learn through engagement, rather than either romantic place attachment or cynical rejection of the community as a legitimate social institution. Attempts to advance our thinking and our actions in the public domain can reveal what we value in the city. This pragmatic reorientation involves a close inspection and consideration of the values of authenticity and individualism (Chapter 4), empowerment, community-building, and social learning (Chapter 5), and risk and resilience (Chapter 6). These values are not selected from above as *a priori* essentials for sustainable and just living. They emerge, instead, from looking at the contemporary urban condition with a pragmatic view and seeking aspects of individual and social identity that hold some power and sway. If we isolate these key aspects and consider their power together, they could constitute a new, uniquely urban understanding of value. In the language of pragmatic sociology, these aspects could constitute a distinctively urban regime of engagement.

## The urban value of authenticity

Much has been made of the individualistic bent of the contemporary urban moment. Individual authenticity and the pursuit of individual self-fulfillment are contemporary urban values par excellence, driven possibly to the point of obsession. There is nothing that particularly weighs toward justice or sustainability in this valuation. At the same time, there may be opportunities for this vision of urban authenticity to be transformed into a pursuit of justice and sustainability. Because individual authenticity in the city takes place in a context of other diverse individuals, each of whom can be expected to hold diverse life plans and pathways, pluralism is valued by this individualism. The notion of multiple, conflicting, perhaps incommensurable ideas of justice is also constant. When people do come together in the urban public sphere, resolving disputes that arise depends heavily upon an ability to construct arguments and justifications that make sense within an individualistic regime of engagement. For a compromise or resolution to be sustainable, it must serve people's felt needs for authenticity and reward those who are successful, self-actualizing life planners. In this familiar mode of engagement, oriented around individual authenticity, traditional liberal understandings of justifications that matter for the resolution of public disputes do not compute. Today's cities are not "formatted" for this kind of argument to hold sway; public conversations are not "equipped" for them. Citizens can lament the loss of civic virtue on the public domain. Urban scholars can critique it. Still, pragmatically speaking, we all need to move on.

It may seem like a dampening of ideals to abandon the old pursuit of a single Great Society in favour of a multiform Great Community of diverse individuals. At the same time, this shifting and localizing of terms and concepts of value suggest a richer understanding of the mix of intimate, personal, social, cultural, economic, and ecological aspects of any person's authentic wellbeing. In the dominant regime of engagement, connecting into the individual's need for authenticity, personal freedom, and the pursuit of wellbeing is essential for a claim to hold water. Neighbourhood-scale work to create and maintain the urban commons can, on the one hand, make more obvious the connection between individual work invested and local group benefit captured. On the other hand, this investment of energy in community work at the local scale can create new habits of reciprocity, collective abundance, authentic experience, leisure, preservation, self-care, local political engagement and dialogue, and other values that fall entirely outside the logic of capitalism. They may well scale up to larger groups and broader goals.

With this in mind, a virtue that we do see exercised in contemporary cities, relevant to the practice of authenticity, is that people's sense of "personality" has become unstuck from their "identity." This is to say that the urban vision of authenticity does not depend upon one's internal consistency, to uphold a particular intransigent identity in all contexts, or else risk losing one's sense of self. To be authentic, urbanites do not need to be stuck within a solitary, unchangeable identity; indeed, to hold oneself to this would be profoundly limiting. This increases the potential for resolution of disputes and the possibility of authentically felt compromises. This kind of compromise is an essential component of progress in the pragmatic way of thinking, as a context-specific mediation and resolution of claims and justifications. It is hampered by those who stick too fiercely to fundamental beliefs and values, out of context. It may be helped by those who are willing to see the possibility for further self-realization through adjusting their views, and so to "change ourselves by changing the city."

Design matters at this interface of individual-urban change. The dynamics of argumentation, or the way in which public disputes play out, and the compromises that are possible and achieved, depend not only upon the mix of interlocutors and their speech conditions, but also upon the material setting in which they interact. Here, the possibilities for the pursuit of urban sustainability and justice need to be considered in the context of the rising orthodoxy of what we can call "placemade" urban design: the kind of mixed-use, middling-density, human-scaled, wide-sidewalked, pocket-parked, transit-connected neighbour-hoods that are the focus of gentrification and revitalization efforts in cities, almost worldwide. As placemaking becomes a nearly universal urban design ideal, it approaches universal "imageability," and this is proving a very strong attractor of those who aspire to situate their authentic identity in the city. These are places that appear open for the development of a human rootedness in the city. We need to consider the questions that are silenced in this process of growing imageability of the placemade city, however. Namely: who are the people left out of this universally-aspiring image and ideal of the urban environment? And, as this city approaches universal imageability, how does it avoid losing its claim to authenticity? Attempts to answer these questions need to be tested from a justice perspective, and in terms of what habits important to the possibility of pursuing justice in the public sphere could be lost, the farther we travel down this particular path.

## The urban value of engagement

For this pragmatic engagement in seeking new compromises to function in the direction of justice and sustainability, not just physical design but also institutional context is important. Despite the rising value and practice of public participation in many domains of urban life, deliberative processes and practices may come to pass without any concomitant emphasis on the merit of socially meaningful outcomes. To seize the potential within this landscape to articulate authentic urban lifestyles and seek a compromise within today's cities, new means of inclusion, new modes of participation, new demands for diversity, intersecting voices, and a complex sense of equality are needed. If this can be achieved, the potential exists to seek a "grammar of commonality through affinity" via participation in public spaces.

Pursuit of the Great Community may still narrow the exclusionary practices of protective social groups, and efforts to create new public spaces may still approximate private enclaves, antagonistic to outside groups and to surprise. Authentic engagement may still leave participants misunderstood and mis-represented. The rich diversity of human experience offers no guarantees and, even in the best case scenario, success in community-building work at the neigh-bourhood scale can only be considered contingent and fragile. At the same time, neighbourhoods, communities, public spaces, and engagement processes need to be created, and existing positive practices need to be supported and advanced. The practice of placemaking in urban neighbourhoods offers a potential union of individualism and social quality. This is key to the pragmatic emphasis on material as well as social and communicative context for debate. In these urban places, a focus on ultra-local experiments and social innovations, opportunities for aesthetic disruptions, the experience of plenitude in public spaces, and open-minded spaces all propel this potential forward.

Forward toward what? Chapter 5 unpacked the pragmatic idea of Utopia. A pragmatic utopia represents not a fully-formed future but a marriage of science and vision that is process-based, indeterminate, and contingent. Coming to rea-lize a pragmatic utopia in the contemporary city depends upon an empowered local population, who evince a willingness to engage in learning together. Seeking a pragmatic utopia entails a shift in aspirations from grand ideals such as the classical liberal model of the Great Society – which, after all, may never support the kind of authenticity and freedom that pluralism requires – to that of Dewey's Great Community. Urbanites make advances toward a vision of the Great Community when they bring their personality and sense of authenticity into the fray, seek new compromises and justifications, and seek to create in the collective. This vision appeals to what Dewey (1927: 149) called "the clear consciousness of communal life."

The vision demands a receptiveness to social learning. Such a commitment is key to the act of creation in a social context, from an interaction between two individuals up in scale to the social sphere of the city as a "learning ecology." Commitment to social learning requires adequate attention to creating a deliberative democratic environment, the physical, social and political space in which discourse and learning can occur, over a sustained period of time. This demand for effective, enduring social learning respectful of the diversity of all is a far cry from the typical way in which participatory processes are pursued: procedurally, with a view to minimizing disagreement, and reaching a kind of consensus as quickly as possible, so as not to prolong discomfort. Instead, a pragmatic view of social learning is inclusive of all possible participants. It also encompasses a much wider range of interactions, from the intimate to the functional and accidental. Individuals conflict, within themselves and between themselves. Communities conflict too. All this is part of the pragmatic appropriation of empowerment.

## The urban value of risk and resilience

When we consider the value attributed to risk and resilience in the contemporary city, a new kind of urban challenge and opportunity reveals itself. Social theorists lean upon Ulrich Beck's theory of the risk society, a view of the societal condition that arose in the 1980s and 1990s, in which the fear of pervasive, uninsurable risk, coming from a multitude of directions, has taken over. Urbanists might claim that this sense of risk had pervaded cities long before it ever became a phenomenon at the scale of the nation state. In the city, the absence of a state of chaos has often seemed a surprise, and much of the thrust of modern urban policy has been to create order and some level of predictability within the tumult of urban action. Some significant portion of the success of the placemaking phenomenon has to do with this very sense of design intervention to create order within urbanity. And people have come, demonstrating the appeal of the additional dose of security, within the general societal preference for risk avoidance and security. To talk about pursuing risk as a virtue, as we do here, seems anachronistic. But upon closer inspection, where placemade urbanism has been particularly successful, it has occurred at the district or neighbourhood scale, bordering existing, older, more chaotic pieces of city; or interspersed within the older layers of less well-designed, more happenstance urbanism. The most successfully urban sites within these districts are the metro stations, the bicycle and walking pathways, and other places that bring the rest of the city into the neighbourhood. It is these open, public spaces of mixing, and the welcome that they offer to the rest of the city to come and spend some time, that offer the most attraction.

From this we can conclude that a very real part of the contemporary appeal of the city, the value that we attribute to urbanism, is the risk that inheres within it. One could argue that this risk is not essential: the city would still have appeal if completely securitized; it would still provide just enough reward. Doubtless, many value the city purely for this purpose. Just as doubtless, much value is created in the city by these tendencies alone. However, in examining the values that go beyond this sense of expediency, we are forced to conclude that the essential experience of urbanism so sought after in the urban celebration today cannot be reduced to expediency. Contemporary urbanism, at least in its ideal form, offers a revaluation of risk and resilience. The rise of urban resilience values is testament to this growing appreciation of the importance of knowledge being able to respond effectively to risk.

Moreover, this attribution of value to urban risk is entirely consistent with the value of authenticity, in multiple forms and expressions; the value of empowerment in the urban public sphere; and the value of work and social learning in the public sphere. Expressing one's personality in authentic ways, raising one's voice, seeking to make public our aspirations and being open to learning about the aspirations of others in our community, even to adjusting our aspirations on this basis – these are risky affairs. In terms of pragmatic sociology, these risks cannot be avoided within the "scene of struggle" of social and political interaction. From engagement in this scene of struggle emerges common civic values. This type of compromise represents the attainment of understanding amidst the non-reconcilable diversity of perspectives and regimes of engagement.

There persists a tension here between the value of risk and the value of security, a value perhaps captured best by the concept of urban resilience. Urban resilience offers an alternative to dealing with risk beyond reliance on insurance. Resilience offers a contingent view of risk, more socially contextualized than actuarial tables, in which the felt impact of a calamity depends on the social response as well as the scale of the calamity itself. There is a push and pull to determining the future in resilience thinking. This is captured best, perhaps, by the notion of a community that "bounces forward" following stress or disaster, rather than simply "bouncing back" to its pre-shock state.

If we recognize the city as a place well suited to "bouncing forward" following adversity, toward new levels of social organization, the utopia of the Great Community is within reach. We can formulate the role of the public sphere in our cities in terms of a place to produce critical social goods, the civilizing power of plurality and diversity, and new civic and urban values at the intersection point of shared interests. This recognition of the need to pursue resilience, in spite of the risk, provides the best defense against the securitization of the city, the "super-gentrification" which is already taking over in the largest, most successfully placemade cities. Insofar as these strategies diminish the pluralism of the city, the availability of public spaces of social interaction, and the social opportunity for engagement with risk as part of conviviality, they eliminate a recognized, critical urban quality, the urban quality of resilience.

## An urban model of justice and sustainability

This book has offered an explanation of social values that exist, to a greater or lesser degree, within the hearts and minds of urbanites today, and which also hold the potential, possibly a uniquely urban potential, to advance justice and sustainability. First, a new understanding of authenticity has been offered. This authenticity springs from fostering and making portable a cosmopolitan and

open stance, astride individual freedoms of mobility and global goals of correspondence, fitting in, relating to the collective. There is also a new understanding of a particularly urban empowerment, of individuals seeking ideals
without any vestige of idealism; that is, a yearning for social transformation
with a disdain for utopian ideals. This kind of empowerment finds its most
natural home at the scale of the community and within the context of open-
minded public spaces. And, finally, there is a new understanding of risk and
resilience, the thrill of the unknown and one's ability to respond and advance
in the face of it. An urban attachment to risk and resilience values diversity and
flexibility in individual achievement and group attachment more than safety
and security. From these values we get: a way to stand up against the social ruin
of extreme gentrification without having to reject urban development outright;
a direction to work in the service of freedom, diversity, contingency, risk; and a
new bias toward partnership in community. Cities are not inherently moral
places, but they can provide opportunities to consider and develop new shared
ideas of morality and expressions of justice.

This set of moral perquisites, particularly when considered in concert, has the
potential to shake up our thinking, in productive ways. The new urban
authenticity is tied, inextricably it seems, to gentrification. Urban discourse
today offers no obvious counter-force to the cultural value of moving-up and
cleaning-up in today's cities. But the urban practice of social risk taking, if
raised to the status of a particularly urban value, offers hope of escaping this
bind. Where cities draw out the energy and potential of human beings, they do
so through offering a resilience-to-risk equation that bucks the pressure of
Joseph's "community trap." It is as tantalizing to those looking for love as those
looking for money (perhaps doubly for those looking for both). Without social,
cultural and economic diversity, the risk of making moves in the city disin-
tegrates, and with this loss goes the reward that comes from surviving those
risks. So, to the extent that capitalist expansion and gentrification processes
destroy risk, they destroy our cities; but this raises another problem. Risky cities
may well continue to attract resilient visionaries and dreamers, but all this risk-
seeking behaviour is atomizing, and ultimately can endanger the social contract.
Here, some contemporary urban trends offer the prospect of a new form of
empowerment that compels urbanites to rise above their self-maximizing
behaviour to envision new social ideals for cities – a new answer to Cruikshank's (1999) "empowerment trap." What this offers is a path to Utopian
thinking about alternative futures that is only justified in terms of practical
measures; a path to shifting values without compelling people to change their
values. They just have to take instrumental action. Maybe, just maybe, this path
to value change feeds into a new sense of authenticity.

The piecemeal and practical-every-step-of-the-way approach to the new
urban transformation is no ideal social scenario, but from a pragmatic view, any
idyllic neo-communitarianism (Davies 2012) seems a long way off. What this
way offers is an understanding of progress toward the just city that embraces the
importance of meaningful process as well as just outcomes. One that is not too

idealistic to gather momentum in an urban population already overburdened by a sense of risk, but one that introduces some insidious ideals that are incontrovertibly different from dominant ones, too. There may be a way within local experiments to commit to radical openness without falling into Purcell's (2006) "local trap," but by producing meaningfully different outcomes in behaviours and habits today, this weekend, and next fiscal quarter.

Allow me to make a personal analogy as a way of defending the difference that is made by the prospect of justice as emergent. I was born into a family of pet lovers and tobacco smokers. One to whom a house without furry friends was quite simply not a home, and a smoke was a necessary accessory and means of punctuating most daily activities. These things constituted the rhythm of our family life, my mother, father, brother, sister and I. Neither of these deep-seated contextual realities seemed to pose any challenge to my healthy brother and sister. I was a different story. As an infant, I had whooping cough, then developed asthma and a persistent skin allergy. I couldn't run for lack of breath. Couldn't be in the same room with smokers. Wasn't comfortable in my skin. My parents, who were kind and caring, who made large sacrifices throughout their lives to ensure opportunity and happiness for their children, and who were certainly dedicated to fairness and justice between their three children, focused on improving the outcomes. Justice, for me, in terms of outcomes, translated into many doctor visits, a regimen of Ventolin and cortisone of various strengths and delivery methods, and an array of solitary pursuits. I could not run, so I was given the opportunity to take music lessons, and had the chance to learn French, while my brother and sister became star athletes.

It wasn't until I had grown up and made my own home, without tobacco smoke or pet hair, that I understood that these lung and skin conditions were environmentally-determined. The whims of genetics, my family perceived, perhaps still believes. But can anyone say, from an outcome-based perspective, that between my brother and sister and I, equity was not achieved? We were able to pursue activities that gave us direction and developed at least some of our respective potential.

Looking at my family from an emergent process perspective, the picture is different. What if somebody had thought about serving justice in my household by removing the smoke and furry pets from my home environment – to meet my right as a child to fill my lungs, run with the other children, compete in sports with my brother and sister and spend long evenings together with family at holiday gatherings? This sounds like something that could have profoundly improved my childhood, at least in my imagination. It would have also changed my family dynamic. Maybe in order to keep Fido and the tobacco addiction, and repair my health, I would have been sent away to boarding school. (How would they have paid for it? Would that have been fair to my brother and sister? Would they have developed a better bond with one another as a result, leaving me out?) Maybe we would have built outdoor cat and dog houses. (Would I have become a competent carpenter? Would my father have lost a finger from a wild power saw?) Maybe I wouldn't have spent so much time reading and would not be writing this book right now. Instead, maybe I would have become a star swimmer or tennis player. Maybe my parents would have quit smoking. Maybe my father would not have died of cancer at age 58. Maybe my mother would not suffer from chronic pneumonia today. Maybe the diseases that lie in store for me, in my own future, would be different ones.

At any rate, while it may seem more immediately, instrumentally practical to right wrongs by focusing on just outcomes, only if we have the courage to consider justice as wildly contingent and totally up in the air, do we have the prospect of advancing our understandings of what justice really ever means in the specific circumstances of our intertwined and largely unpredictable lives.

The opportunities that could lead to a uniquely urban moral thinking are themselves precarious and fraught. Authenticity, empowerment, and risk are not meaningful conduits to moral thinking if they can only be enjoyed by people of means and in a tokenistic way. We must not lose sight of the need for critique, even as we avoid a nihilism that offers only abject resignation. The new urban morality demands that we rest neither too comfortably with ourselves nor too comfortably with the city. What justice demands has yet to be articulated in the wealth of dialogue and literature specific to the city in our time; but Fainstein (2010) offers an excellent guide. It demands, as Friedmann (2002) knew, holding opposing ideas in tension. Opposing ideas such as: cities that enable the balancing of a drive for uniqueness and autonomy at the individual level with a desire for institutions looking out for the good of the community. Opposing ideas such as: upholding the attainment of individual happiness, and privileging the common good. Opposing ideas such as: permitting pragmatic action as well as the creation of utopian ideals.

## Passion for the city is back. So what?

Passion for the city is back. If we who dwell in cities hope to find distinctively urban value, we have an interest in keeping alive an active critique of the urban celebration in which we engage. We have a common interest in remaining curious in the face of this passion. If we can do this, we have a hope of finding enough material difference in urban action to sustain the city against the hypocrisy of the current urban celebration, and to move toward an essentially urban, moral platform on which to build a social and sustainable future.

The contemporary urban celebration suggests the end of the road for the older understanding of the geography of injustice that sees the city as a stand-in for extreme social ills, the way it was in Banfield's day. In order to advocate for the role of the city in further articulating possibilities for justice, the task before willing urbanists is neither to dismiss nor to celebrate the implications of the new urban geography that we do not yet understand. As soon as we are certain that the old geography of injustice has disappeared, we can be certain that a new geography of injustice has replaced it. Where does injustice concentrate, in the new urban scene? How can we recognize it? How can we argue against it? Build the capacity to address it?

What I have suggested is that a case exists for moral thinking about the city and its role in human development. It's there, in a tentative, piecemeal and contingent way; embodied in the experiences of urbanites, and in the thinking of some of the city's most thoughtful advocates. As urbanists, we ought to be skeptical of the new enthusiasm for the city; but we are naturally curious about the city's potential. Out of an urban celebration that often sweeps away difficult questions, and resists being substantiated, cities have an immanent morality worthy of attention and development. And, while acknowledging critique that warns us of various traps that exist in a focus on the city, a pragmatic approach leads us to conclude that a uniquely urban morality can lead us toward more just processes and outcomes. Urban life is exciting: today's chorus of urban celebration reveals just how exciting the city appears across many domains of thinking, from economics, to sociology, to political science, to design and engineering. While there is more gloss than proof in some of these arguments, they are too numerous to ignore completely. Moreover, our experientially shared excitement about the city, cutting across these other celebrations, finds common roots in what we call three essentially and singularly urban features: authenticity, empowerment, and resilience. Just the same, as much as the pursuit of authenticity, empowerment, and resilience is urban, this pursuit is not necessarily moral. Between uncritical urban celebration and despair-inducing critique, is this pragmatic argument for passion, i.e., facing up to the worth of cities in generating a swell of moral thought from the ground that we are given.

A pragmatic approach is frequently accused of being Polyanna-ish, naïve, and excessively optimistic about the case for hope in human, social change. A more accurate reading of the optimism within the philosophy of pragmatism is that it is too doggedly humanistic to abandon hope in human-driven progress. On top of this basic, anti-foundational humanist and evolutionary approach of pragmatic philosophy, the argument in this book adds the methodological value from the pragmatic sociology of critique. The analytical strength of a critical pragmatic approach to understanding the structure and function of social disputes adds to the sense of realism within pragmatism. It does this by inserting an understanding of the tools that people use in discourse in social settings in order to argue, test one another's beliefs and actions, struggle toward compromises, and form new values. The goal, in sum, of a pragmatic approach, is to allow us to "get down to the nitty-gritty, how we are to decide what is right and wrong, and how we are to figure out which value judgments are true and false in specific situations" (Bernstein 2010: 167). Those who cannot advance their arguments based on a suitable regime of engagement, with the kinds of justifications that are recognized by others, in the right circumstances for their voices to be amplified and carried forward, are likely to be ignored. And it is analytically unhelpful to associate particular individuals and groups with particular positions and actions. Instead, people change their minds and behaviours, for many different reasons, one of which is that certain situations demand certain kinds of behavioural norms. This ability to adapt one's views to the demands of a situation ought to be valued as key to authentic urban engagement: "To adopt the attitude called for by the situation is to become a being belonging to the world from which the situation arises (when one adopts the attitude appropriate to a voting booth, for instance, one becomes a citizen)" (Boltanski and Thévenot 2006: 147). Sorting out people's responses and reasoning surrounding their actions in the face of competing normative claims is key to what Bernstein (2010: 167) calls the work of a "good pragmatist." Bernstein explains what this work entails:

> The really hard moral and political issues concern just how we are to figure out what is to be done and how we are to judge competing claims ... how we are to resolve the type of conflicts that constantly arise in a democratic society – especially when reasonable citizens sharply disagree over fundamental issues.
>
> (ibid.: 167)

None of this is to deny that substantial and significant constraints exist on the potential to pursue justice and sustainability in cities. Seeking to construct, maintain, and spread, let alone win, arguments about justice and sustainability will continue to be an uphill battle. It always has been. At the same time, it is precisely up to those committed to the city, to force new conversations about the need to address justice as part of becoming more resilient to a sometimes devastating political and cultural climate. This is not a call for consensus. It *is* a call for willingness to compromise. Nor is it really a call for cooperation, although it *is* a call to create a common domain for community work. It *is* a call to value the authenticity, empowerment, community, risk and resilience within a cacophony of different ideas about how to move forward at all. This cacophony may be mere noise at formal scales of human societal organization – but it is the buzz of potential in the city. This mode of engagement and style of action were viewed by John Dewey a century ago as providing a sense of the better life to be lived:

> As the new ideas find adequate expression in social life, they will be absorbed into a moral background, and will the ideas and beliefs themselves be deepened and be unconsciously transmitted and sustained. They will color the imagination and temper the desires and affections. They will not form a set of ideas to be expounded, reasoned out and argumentatively supported, but will be a spontaneous way of envisioning life.
>
> (2004 [1920]: 121)

In the thought experiment that opened this book, I misrepresented the street protest in Quebec City as a hypothetical case. In fact, this case, which set the stage for the argument developed here, was a personal experience, from early 2001, that set the course of my thinking in this direction. I changed a few details. But I was there too, on those cold and narrow fortress streets, head full of righteous indignation against injustices, global and local, voice hoarse from all the collective chanting, nose crusted with blood from the tear gas, ears ringing from the sheer diversity of versions of justice that I was bumping into. Many of you readers have been there, too, setting your sights in some way on ordinary justice in the situations you are given. Are we on the right side of the fence? It is a start.

# References

Banfield, Edward. *The Unheavenly City*. Boston: Little, Brown, 1968.

Bernstein, Richard. *The Pragmatic Turn*. Malden, MA: Polity Press, 2010.

Boltanski, Luc and Laurent Thévenot. *On Justification: Economies of Worth*. Princeton, NJ: Princeton University Press, 2006 [1991].

Castells, Manuel. *The Urban Question*. Cambridge, MA: MIT Press, 1977.

Cruikshank, Barbara. *The Will to Empower: Democratic Citizens and Other Subjects*. Ithaca, NY: Cornell University Press, 1999.

Davies, William. "The Emerging Neocommunitarianism." *The Political Quarterly* 83(4) (2012): 767–776.

De Angelis, Massimo. "Does Capital Need a Commons Fix?" *Ephemera: Theory & Politics in Organization* 13(3) (2013): 603–615.

Dewey, John. *Reconstruction in Philosophy*. Mineola, NY: Dover Press, 2004 [1920].

Dewey, John. *The Public and its Problems*. New York: Henry Holt & Co., 1927.

Fainstein, Susan. *The Just City*. Ithaca, NY: Cornell University Press, 2010.

Friedmann, John. *The Prospect of Cities*. Minneapolis, MN: University of Minnesota Press, 2002.

Gibson-Graham, J.K. *A Post-Capitalist Politics*. Minneapolis, MN: University of Minneapolis Press, 2006.

Hoch, Charles. "Pragmatic Communicative Action Theory." *Journal of Planning Education and Research* 26 (2007): 272–283.

Parnell, Susan and Jennifer Robinson. "(Re)theorizing Cities from the Global South: Looking Beyond Neoliberalism." *Urban Geography* 33(4) (2012): 593–617.

Purcell, Mark. "Urban Democracy and the Local Trap." *Urban Studies* 43(11) (2006): 1921–1941.

Standing, Guy. *The Precariat*. London: Bloomsbury Academic, 2011.

Watson, Vanessa. "Seeing from the South: Refocusing Urban Planning on the Globe's Central Urban Issues." *Urban Studies* 46(11) (2009): 2259–2275.

Zukin, Sharon. *The Cultures of Cities*. Malden, MA: Blackwell, 1995.

# Index